VOL. 1
Uncovering
Student Ideas
in Physical Science

45 NEW Force and Motion Assessment Probes

VOL.1

Uncovering Student Ideas in Physical Science

45 NEW Force and Motion Assessment Probes

By Page Keeley and Rand Harrington

National Science Teachers Association
Arlington, Virginia

National Science Teachers Association

Claire Reinburg, Director
Jennifer Horak, Managing Editor
Andrew Cooke, Senior Editor
Judy Cusick, Senior Editor
Wendy Rubin, Associate Editor
Amy America, Book Acquisitions Coordinator

ART AND DESIGN
Will Thomas Jr., Director
Cover, Inside Design, and Illustrations by Linda Olliver

PRINTING AND PRODUCTION
Catherine Lorrain, Director

NATIONAL SCIENCE TEACHERS ASSOCIATION
Francis Q. Eberle, PhD, Executive Director
David Beacom, Publisher

Copyright © 2010 by the National Science Teachers Association.
All rights reserved. Printed in the United States of America.
13 12 11 10   5 4 3 2

Library of Congress Cataloging-in-Publication Data
Keeley, Page.
  45 new force and motion assessment probes / by Page Keeley and Rand Harrington.
    p. cm. -- (Uncovering student ideas in physical science ; v. 1)
  Includes bibliographical references and index.
  ISBN 978-1-935155-18-8
  1. Force and energy--Study and teaching. 2. Motion--Study and teaching. 3. Educational evaluation. I. Harrington, Rand. II. Title. III. Title: Forty-five new force and motion assessment probes.
  QC73.6.K44 2010
  530.071--dc22
                2010010354

eISBN 978-1-936137-70-1

NSTA is committed to publishing material that promotes the best in inquiry-based science education. However, conditions of actual use may vary, and the safety procedures and practices described in this book are intended to serve only as a guide. Additional precautionary measures may be required. NSTA and the authors do not warrant or represent that the procedures and practices in this book meet any safety code or standard of federal, state, or local regulations. NSTA and the authors disclaim any liability for personal injury or damage to property arising out of or relating to the use of this book, including any of the recommendations, instructions, or materials contained therein.

PERMISSIONS
You may photocopy, print, or e-mail up to five copies of an NSTA book chapter for personal use only; this does not include display or promotional use. Elementary, middle, and high school teachers *only* may reproduce a single NSTA book chapter for classroom or noncommercial, professional-development use only. For permission to photocopy or use material electronically from this NSTA Press book, please contact the Copyright Clearance Center (CCC) (*www.copyright.com*; 978-750-8400). Please access *www.nsta. org/permissions* for further information about NSTA's rights and permissions policies.

Contents

Foreword by Jim Minstrell, FACET Innovations ix

Preface xi

Acknowledgments xxiii

About the Authors xxv

Introduction 1

Section 1. Describing Motion and Position

    Concept Matrix 12
    Related Curriculum Topic Study Guides 13
    Related NSTA Press Books, NSTA Journal Articles, and
        NSTA Learning Center Resources 13

1   How Far Did It Go? 15
2   Skate Park 19
3   Following Jack: Part 1 23
4   Following Jack: Part 2 27
5   Go-Cart Test Run 31
6   Checking the Speedometer 35
7   Speed Units 39
8   Just Rolling Along 43
9   Crossing the Finish Line 47
10 NASCAR Racing 51
11 Roller Coaster Ride 55
12 Rolling Marbles 59
13 String Around the Earth 63

Section 2. Forces and Newton's Laws

    Concept Matrix 68

    Related Curriculum Topic Study Guides 69

    Related NSTA Press Books, NSTA Journal Articles, and

        NSTA Learning Center Resources 69

14   Talking About Forces 71

15   Does It Have to Touch? 75

16   Force and Motion Ideas 79

17   Friction 83

18   A World Without Friction 87

19   Rolling to a Stop 91

20   Outer Space Push 95

21   Riding in the Parade 99

22   Spaceships 103

23   Apple in a Plane 107

24   Ball on a String 111

25   Why Things Fall 115

26   Pulling on a Spool 119

27   Lifting Buckets 123

28   Finger Strength Contest 127

29   Equal and Opposite 131

30   Riding in a Car 135

Section 3. Mass, Weight, Gravity, and Other Topics

    Concept Matrix 140

    Related Curriculum Topic Study Guides 141

    Related NSTA Press Books, NSTA Journal Articles, and

        NSTA Learning Center Resources 141

31   Pizza Dough 143

32   What Will Happen to the Weight? 149

33   Weighing Water 153

34   Experiencing Gravity 157

35	**Apple on the Ground**	163
36	**Free-Falling Objects**	167
37	**Gravity Rocks!**	171
38	**The Tower Drop**	177
39	**Pulley Size**	181
40	**Rescuing Isabelle**	185
41	**Cutting a Log**	189
42	**Balance Beam**	193
43	**Lifting a Rock**	197
44	**The Swinging Pendulum**	201
45	**Bicycle Gears**	205
	Index	209

Dedication

This book is dedicated to Rand's parents, Bev and Don Harrington, and to Bev Cox, an extraordinary science specialist retired from the Orange County School District in Orlando, Florida. Through Bev's leadership, formative assessment has impacted hundreds of teachers and thousands of students in one of the largest school districts in the nation.

Foreword

Formative assessment—or assessment for learning—has become an increasingly common focus for teachers and schools since the late 1990s. Touted by research as the single most effective strategy for advancing learning for all students, formative assessment has been incorporated by more and more teachers into their classroom practices. As they do so, they are discovering that the process is far more complex than simply administering a probe or checking in occasionally to see if students are "getting it." Teachers are finding that to effectively use assessment results to further learning, the strategies they employ have to be carefully linked to specific learning goals. Formative assessment provides them with rich information that allows them to understand not only *what* students have learned but *how* they are learning. Teachers see the importance of asking students two critical questions: *What do you know?* and *How do you know that?* Getting answers to these questions are especially important in assessing conceptual learning (as opposed to assessing skill mastery) in science and mathematics.

To really attend to students' ideas in the classroom requires a change in perspective for teachers and students. Many of the teachers with whom we have worked describe it as a move from a focus on teacher performance to a focus on student learning—a shift from teacher- or lesson-based learning to learning-based lessons. A focus on student learning also means using probes, elicitations, and lessons that will help teachers answer a question they must pose to themselves: *What and how are my students learning in relation to the learning goal?* This focus also entails reflecting on the information collected and interpreting it to answer two additional questions: *What are the strengths and problematic aspects of the students' thinking?* and *What do these students need next to deepen their learning?* Finally, a focus on student learning means that subsequent instruction addresses the question *What learning experience or feedback will address the identified needs as well as the learning goal?* This cycle of collection, interpretation, and action is an ongoing process in which the art and science of instruction meet. In addition, it requires teachers to draw more heavily than they have before on their science content and pedagogical content knowledge (knowledge of how students learn a particular science concept) in order to make the cycle effective. As many teachers have observed, formative assessment initially appears to be easy to carry out. In fact, that impression is deceptive. Formal assessment takes time, tenacity, and resources if it is to be done well.

Uncovering Student Ideas in Physical Science is the latest addition to an important series of resources that support teachers who are building and refining their use of formative assessment in science. The re-organization of resources around important "big ideas" in science not only emphasizes the targeted nature of the formative assessment process but also is likely to help teachers address their own learning needs more systematically. As in the earlier, four-volume *Uncovering Student Ideas in Science* series by Page Keeley and her colleagues, she and Rand Harrington in this volume provide information and references in the Teacher Notes accompanying each probe to help teachers scaffold their own learning as well as that of their students. The Purpose description, for example, provides important information about what a specific probe is

Foreword

expected to elicit and how this might be related to larger learning goals. This discussion of the purpose, together with the Related Ideas in the National Standards sections, will help teachers be more purposeful in collecting information about student learning and in selecting assessment tools and strategies that are closely tied to a specific learning goal.

The suggestions provided in Administering the Probe can be used by teachers to anticipate and eventually interpret their students' responses. This section and the Related Concepts section will support teachers in their efforts to become more flexible in their assessments and to hone in on student understanding of a particular idea. These sections also underscore the ongoing, targeted nature of formative assessment. For example, the half-dozen probes in the book that are related to Newton's first law could be used at various points in a unit to explore students' developing conceptual understanding. One or two probes could be used early in the unit when focusing on straight-line motion. Another probe could be used midway by relating the first law to relative motion. Finally, three other probes could be used when instruction focuses on the topics of changing direction and circular motion. Alternatively, the same six probes could be used by a school or district to articulate its curriculum regarding motion. As noted in the probes themselves, some would be appropriate at the intermediate or middle school level, while others could be used for review and extension at the high school level.

Both the Explanation and Related Research sections offer resources to support teachers' knowledge needs. The brief scientific explanation provided with each probe not only helps teachers to better interpret their students' responses but also could alert them to their own content-related learning needs before they engage their students. Meanwhile, the descriptions of findings from related research give teachers additional information about the intention of each probe and about why the student thinking it elicits matters in the overall progression of learning. Teachers can use this information to develop the pedagogical content knowledge that is so important to effective formative assessment. In turn, they will be able to better understand how students learn the subject of force and motion.

Although no written resource will provide teachers with a "silver bullet" solution to every challenge, this new *Uncovering Student Ideas in Physical Science* series promises to do more than any other resource I know of to give support to teachers who are trying to engage in effective formative assessment in science. This book and the four volumes to follow in the series on physical science, together with earlier works such as *Science Curriculum Topic Study* and *Science Formative Assessment*, are essential additions to the library of any professional science educator or organization. This growing set of tools provides teachers with greater opportunities to explore their students' thinking, challenge their own, and have fun in the process.

Jim Minstrell
FACET Innovations

Preface

Background

Since the late 1990s, K–12 science educators have acknowledged the critical importance of using diagnostic tools and formative assessments to improve teaching and learning. Significant research-based publications such as *How People Learn: Brain, Mind, Experience, and School* (Bransford, Brown, and Cocking 2000) raised teachers' awareness of the need to identify the preconceptions that students bring to the science classroom and use these preconceptions as springboards for learning. However, student- and teacher-friendly questions that make K–12 students' thinking visible to themselves and their teachers were not readily available. Thus, the four-volume *Uncovering Student Ideas in Science* series was born.

In October 2005, the first book in the series rolled off the press, making its debut at the NSTA Area Conference in Hartford, Connecticut. The book was the first of its kind. It gave K–12 educators a set of probing science questions to use with students on life, Earth, space, and physical science. The "probes" link key ideas in the K–12 content standards (AAAS 1993 and NRC 1996) to research on students' commonly held ideas. Each probe is accompanied by detailed teacher background notes on science content and suggestions for use in the classroom. Although the physics education community had developed diagnostic assessments for many years for high school and undergraduate physics students, such as the Force and Motion Conceptual Evaluation (Thornton and Sokoloff 1998), no similar assessment that probed K–12 students' science preconceptions existed until Volume 1 in the *Uncovering Student Ideas in Science* series was published.

Three additional volumes later—with a total of 100 K–12 formative assessment probes now published—the *Uncovering Student Ideas in Science* series has soared in popularity. The four books are used by thousands of K–12 classroom teachers, university professors, professional developers, instructional coaches and mentors, and even parents. The series has become a valuable resource for improving student learning and deepening teachers' science content knowledge.

Each volume's introduction contains vital information about the use of the probes and formative assessment:

- Introduction to Vol. 1: Presents an overview of formative assessment: what it is and how it differs from summative assessment. Background on probes as specific types of formative assessments and how they are developed is provided.
- Introduction to Vol. 2: Describes the link between formative assessment and instruction and suggests ways to embed the probes into your teaching.
- Introduction to Vol. 3: Describes how to use the probes and student work—either individually or through professional learning communities—to (a) deepen your understanding of students' ideas and the implications of those ideas for instruction, (b) learn new science content, or (c) even uncover a deeply rooted misconception you might have.
- Introduction to Vol. 4: Describes the link between formative and summative assessment. Gives reasons why an investment in formative assessment before and throughout instruction can improve students' performance on the summative end.

Preface

Collectively, these four introductions will expand your assessment and instructional repertoires and will deepen your understanding of students' thinking and effective science teaching.

The New Series: *Uncovering Student Ideas in Physical Science*

In 2009, after the publication of Volume 4, Page Keeley, principal author and probe developer of the *Uncovering Student Ideas in Science* series, decided to begin a new set of books that would focus on specific topics in physical science. The new series is called *Uncovering Student Ideas in Physical Science*. This book—on force and motion—is the first in that anticipated five-volume series. The topics of the four volumes to come will be electricity and magnetism, energy, sound and light, and matter.

Format of This Book

This book contains 45 probes on force and motion for a variety of grade levels, from elementary grades up through high school. It is made up of three sections: Section 1, Describing Motion and Position; Section 2, Forces and Newton's Laws; and Section 3, Mass, Weight, Gravity, and Other Topics. The format is similar to that of the earlier four volumes, with a few changes due to the topic-focused nature of this series. For example, the book's introduction (starting on p. 1) focuses specifically on why force and motion ideas are challenging to teach and learn. As in the earlier series, each probe is followed by a Teacher Notes section, which is made up of these eight elements.

1. Purpose

This section describes the purpose of the probe—that is, why you would want to use it with your students. It also states the general science concept or topic targeted by the probe and the specific ideas the probe is trying to elicit from students. Before using a probe, you should be clear about what the probe can reveal. Taking time to read the purpose will help you decide if the probe fits your intended use.

2. Related Concepts

Each probe is designed to target one or more related concepts that often cut across grade spans. These concepts are also included on the concept matrix charts on pages 12, 67, and 140. A single concept may be addressed by multiple probes as indicated on the concept matrix. You may find it useful to use a cluster of probes to target a concept or specific ideas within a concept. For example, there are six probes that relate to Newton's first law of motion.

3. Explanation

A brief scientific explanation accompanies each probe and clarifies the scientific content that underlies the probe. The explanations are designed to help you identify what the "best" or most scientifically acceptable answers are (sometimes there is an "it depends" answer) as well as clarify any misunderstandings you might have about the content.

The explanations are not intended to provide detailed background knowledge about the content nor are they written for an audience of physicists. In writing these explanations, we were careful to keep in mind that our audience will include teachers who have little or no physics background. We tried not to oversimplify the science but at the same time we wanted to provide the science content a novice teacher would need. If you need additional background information on the content, refer to the NSTA resources listed at the beginning of each section (pp. 13, 69–70, and 141–142), such as *Force and Motion: Stop Faking It! Finally Understanding Science So You Can Teach It* (Robertson 2002) or a Science Object from the NSTA Learning Center.

Preface

4. Administering the Probe

In this section, we suggest ways to administer the probe to students, including appropriate grade spans and, in some cases, modifications to make a probe that is intended for one grade span useful for another. A modification might be to eliminate certain choices from the list of possible answers for younger students, who may not be familiar with particular words or examples. In other probes, we might recommend using the word *weight* instead of *mass* as a "stepping-stone" concept with younger elementary students, who frequently confuse the word *mass* with the phonetically similar word *massive*. (See a description of stepping-stone ideas on p. 5 in the introduction.) The Administering the Probe section also tells teachers when it is a good idea to establish the context for the probe by showing students the various items that are referred to in the probe (e.g., a ramp, a ball). Also, although the probes were designed to be used by individual students, this section occasionally recommends that a probe be used in small groups.

The suggested grade levels, which appear in the concept matrices that precede each section, are intended to be suggestions only. Your decision about whether or not to use a probe depends on why you are using the probe and on your students' readiness. Do you want to know about the ideas your students are expected to learn in your grade-level standards? Are you interested in how preconceived ideas develop and change across multiple grade levels in your school whether or not they are formally taught? Are you interested in whether students achieved a scientific understanding of previous grade-level ideas before you introduce higher-level concepts? Do you have students who are ready for advanced concepts? We recommend that you weigh the suggested grade levels against the knowledge you have of your own students, your school's curriculum, and your state's standards.

5. Related Ideas in the National Standards

This section lists the learning goals stated in the two national documents generally considered the national science standards: *Benchmarks for Science Literacy* (AAAS 1993) and the *National Science Education Standards* (NRC 1996). The learning goals from these two documents are quoted here because almost all state standards are based on them. Also, because the probes are not designed as summative assessments, the learning goals in this section are not intended to be considered alignments but rather ideas that are related in some way to the probe.

Some targeted probe ideas, such as the simple machines–related ideas in Section 3, are not explicitly stated as learning goals in the standards, but they are clearly related to concepts in the standards that address specific ideas about forces and energy. The national science standards do not include simple machines for the purpose of learning the types and names of simple machines; instead, simple machines are used as a context in the standards for important learning goals. However, because ideas such as simple machines often appear in state standards and are commonly used in curriculum materials, we have included them here as "Other Topics" (probes #39–#45). When the ideas elicited by a probe appear to be a strong match (aligned) with a national standard's learning goal, these matches are indicated by a star symbol (★). You may find these "matches" helpful when you are using probes with lessons and instructional materials that are strongly aligned to your state and local standards and specific grade level.

Sometimes you will notice that an elementary learning goal from the national standards is included in middle school and high school probes. This is because it is useful to see the related idea that the probe may build on from a previous grade span. Likewise, sometimes a high school learning goal from the national

Preface

standards is included even though the probe is designated for grades K–8. This is because it is useful to consider the next level of sophistication that students will encounter in their spiraled learning.

6. Related Research

Each probe is informed by research regarding students' typical preconceptions about force and motion. The authors primarily drew on two comprehensive research summaries commonly available to educators: Chapter 15 in *Benchmarks for Science Literacy* (AAAS 1993) and Driver's *Making Sense of Secondary Science: Research Into Children's Ideas* (Driver et al. 1994). We also examined physics education research. The research findings will help you better understand the intent of the probes and the kinds of thinking your students are likely to reveal when they respond to them. We encourage you to seek new and additional published research. The Curriculum Topic Study (Keeley 2005) website at *www.curriculumtopicstudy.org* has a searchable database to access additional current research articles on learning.

It should be noted that although many of the research studies we cite were conducted in past decades and involved children in other countries as well as the United States, most of the results of these studies are considered timeless and universal. Whether students develop their ideas in the United States or in other countries, research indicates that many of their ideas about force and motion are pervasive, regardless of geographic boundaries and societal and cultural influences.

7. Suggestions for Instruction and Assessment

A probe remains simply diagnostic, not formative, unless you use the information acquired from your students to inform your instruction. After analyzing your students' responses, the most important step is to decide on the student interventions and instructional paths that would work best for you, based on your students' thinking. We have included suggestions gathered from the wisdom of teachers, the knowledge base on effective science teaching, research, and our own collective experiences working with students and teachers. In the "Suggestions for Instruction and Assessment" section, you will not find extensive lists of detailed instructional strategies but rather brief suggestions for planning or modifying your curriculum or instruction to help students learn ideas that they may be struggling with. After administering a probe, you may find that you need to provide an effective context for certain topics or that using a bridging analogy would work for you and your students. Effective contexts and bridging analogies are among the suggestions in this section.

Learning is a very complex process and no single suggestion will help all students learn. But formative assessment encourages you to think carefully about the variety of instructional strategies and experiences needed to help students learn scientific ideas. As you become more familiar with the ideas your students have and the many factors that may have contributed to their misunderstandings, you will identify additional strategies to teach for conceptual change.

8. References

References are provided for the standards, research findings, and some of the suggestions cited in the Teacher Notes. You might want to read the full research summary or access a copy of the research paper or resource cited in the Related Research section.

Introductory Elements of Each Section

Each of this book's three sections begins with three elements.

1. **Concept Matrix** (pp. 12, 68, and 140).

Preface

The matrix allows the user to see at a glance what science concepts are related to each probe and what the most appropriate grade level is for using the probe.

2. **Related Curriculum Topic Study (CTS) Guides.** This is a list of four or five topics that are covered in the book *Science Curriculum Topic Study: Bridging the Gap Between Standards and Practice* (Keeley 2005) and are related to the probes in that section. The guides in *Science Curriculum Topic Study* were used to inform the development of the probes in this book. For more information on how the guides themselves were developed, and an example of a guide and how it was used to develop a probe, see the Appendix on pages xix–xxi.

3. **Related NSTA Press Books, NSTA Journal Articles, and NSTA Learning Center Resources.** NSTA's journals and books are increasingly targeting the ideas that students bring to their learning. We have provided a few suggestions for additional readings and online resources that complement or extend the use of the individual probes and the background information that accompanies each probe. These resources can be used to improve teachers' content knowledge, pedagogical content knowledge, and instructional repertoire. Searching the Learning Center on the NSTA website at *http://learningcenter.nsta.org/?lid=hp* will provide additional resources.

Using the Probes: Curricular and Instructional Considerations

Unlike most summative assessments, the probes in this book are not limited to one grade level. What makes them interesting and useful is that they provide insights into the knowledge and thinking students may acquire as they progress from one grade span to the next. Sometimes the same idea comes up at different grade levels in a different context. Teachers in the middle school or high school grades might ask teachers in the elementary grades to give a probe to younger students, who may already have ideas about a particular concept before they formally encounter the concept in later grades. Elementary teachers can share the results with middle school or high school teachers so those teachers learn the kinds of ideas younger students have already formed that may affect their learning of more sophisticated concepts in later grades. Some probes can be used across elementary, middle school, and high school grades; others may cross over just a few grade levels; and a few are designed specifically for high school physics students or for elementary students. Teachers from different grade spans (e.g., in middle school and high school) with a spiraling curriculum could administer the same probe and come together and discuss their findings. To do this, it is helpful to know what students typically experience at a given grade span about force and motion. The instructional and curricular considerations described in the essays about force and motion in *Benchmarks for Science Literacy* (AAAS 1993) and the *National Science Education Standards* (NRC 1996) are summarized below.

Elementary Grades

During the early grades—from preschool and kindergarten through grade 2—children should have many varied opportunities to observe and describe all kinds of moving things. They should be encouraged to ask questions such as, Does it [an object] move in a straight line? Does it move fast or slow? How far will it move? What direction will it move in? Where does its motion start? Where does its motion end? How can I make it move

Preface

(including using magnets to make things move without touching)? How can I change the way it moves? Teachers should encourage the children to investigate their questions by recording and drawing what they see.

This is also the time when young children should be developing the descriptive language needed to describe motion and position, using terms such as *straight, zigzag, round and round, back and forth, fast and slow, up, down, in front,* and *behind.* They should also have opportunities to use simple measurement devices, including ones they create, and be introduced to techniques to measure distance and time.

During the intermediate grades (grades 3–5), students should continue describing motion. They should also sharpen their measurement skills, becoming more quantitative in their measurements. Their descriptions and drawings will become much richer and more detailed than in their early days in school. They should become increasingly familiar with techniques and units for measuring distance, time, and speed. Varied opportunities should be provided to manipulate objects by pushing, pulling, throwing, dropping, sliding, and rolling. Students in grades 3–5 should also become increasingly facile with measurement tools such as rulers, tape measures, clocks, and stopwatches. Recording motion data and looking for patterns on simple grids and graphs should be integral parts of the curriculum. By the end of fifth grade, students should be able to describe speed as the distance traveled in a given unit of time.

This idea that forces can move things at a distance without objects touching is further developed in grades 3–5 as students investigate pushes and pulls using magnets and electrically charged objects. Through these experiences, they begin to work out some of the general relationships between force and a change in motion. They should investigate and experience a variety of forces. During this grade span, most students internalize a force as a push or pull of one thing on another, a prerequisite to the notion of interactions.

Early notions about gravity develop before students are ever introduced to the word. In the early years, they observe that things fall downward if there is nothing to hold them up. In grades 3–5, they develop the notion that things fall or remain on the ground because the Earth *pulls* on them. During these years, students have many experiences balancing objects and begin to recognize quantitative aspects of balancing.

Middle School

Students continue describing motion, with increasing attention to appropriate scientific terminology. Using simple objects, such as rolling balls and mechanical toys, students move from qualitative to quantitative descriptions of motion and describe the forces acting on the moving objects. Given opportunities to see the effect of reducing friction, they can begin to imagine a frictionless situation when describing motion. At this level, students move from thinking about motion in terms of motion-or-no-motion to categorizing motion as steady motion (e.g., constant speed, uniform motion), speeding up, and slowing down. They perform basic calculations to determine speed and recognize the difference between speed in a single moment of time versus average speed.

This is a time to connect students' learning in mathematics to applications in science because a conceptual and procedural understanding of ratio and proportion is important to quantitatively describing motion. In middle school, students transition from thinking about just speed, to speed and direction, using the term *velocity*. (*Note:* Acceleration is a difficult concept at this level and is not emphasized in the grade-level standards.) Students become more skilled and precise in their use of measurement tools and techniques, including the

Preface

use of computer probeware to analyze motion. Representations such as motion diagrams and graphs should be used to encourage students to analyze and communicate motion data.

The relationship between force and motion is further developed at this grade level. Students become familiar with the concept of inertia and seem to have little problem with believing an object at rest tends to stay at rest unless acted upon by an outside force. They can relate everyday phenomena to this idea. However, they have a difficult time accepting the idea that an object in motion will keep moving unless a force is applied to it. The more experiences students have seeing the effect of reducing friction and discussing it, the easier it may be to get them to accept the idea that a moving object will keep moving in a frictionless environment.

Students should now develop the notion of balanced and unbalanced forces and should have ample opportunity to describe the forces acting on objects. Instruction should include opportunities to think about both active and passive forces because students at this age tend to equate force with motion and may think there is no force acting on an object that is not moving, such as a book on a table.

The notion of gravitational force is introduced at this level. Students move beyond terrestrial gravity to realizing that gravity applies to all matter everywhere in the universe. They also move beyond thinking about the Earth *pulling* to the idea of gravity as a force directed from Earth's center. They also begin to learn qualitatively that gravitational force depends on the size of the masses and the distance between them.

Overall, the concrete experiences students have in middle school with identifying forces and describing motion provide the foundation on which a more comprehensive and detailed understanding of force and motion will be developed in high school.

High School

In high school, students learn to use more sophisticated mathematics to represent various motions, a field of study within physics called "kinematics." However, students need explicit help to maintain a connection between these representations and the actual motion. Position, velocity, and acceleration graphs are an integral part of their learning, and students spend time thinking deeply about the differences between a quantity and a change in that quantity. At this level, students realize the power of mathematics in describing and representing real-world phenomena and are expected to be able to calculate and then interpret the slope of a graph. Motion detectors are frequently used to help students learn these concepts. These detectors provide real-time graphs of the motion of an object (such as a walking student) and can be a significant benefit to those who struggle to understand the connection between a graph and the motion.

In addition to graphs, students are also introduced to vectors (arrows) as a way to represent position, velocity, and acceleration. Although this representation can be very useful, vectors can also be difficult to understand because the length of each vector (an arrow) can represent different quantities (such as the speed or the acceleration of the object). At a basic level, students can learn how to add and subtract vectors graphically, while students with the prerequisite mathematics skills and conceptual understanding can learn to use trigonometry to analyze vector components.

An understanding of vectors and motion graphs is then typically applied to objects that are moving in two dimensions, such as projectile or circular motions. One increasingly popular method of collecting data of objects moving in two dimensions is to use video cameras. The short film clips can be imported into a computer and used to measure the position

Preface

of the object as a function of time and to generate motion graphs.

After students have a firm understanding of kinematics, the topics of dynamics are typically introduced. Dynamics includes the study of forces (Newton's three laws), momentum, energy, and rotations. Students learn how to use their understanding of motion to infer the presence of a net force (when there is a change in the velocity of an object) and to infer changes in mechanical energy. Mechanical energy is the combination of kinetic energy (energy of motion) and gravitational potential energy (energy related to the location of the object near other masses like the Earth). Forces that are applied to an object and result in the object moving can cause a change in these energies.

One of the primary problem-solving tools introduced in high school physics courses is the free-body diagram. The purpose of this diagram is to indicate all the forces acting on a single object. Each force on the diagram is represented by an arrow (or vector) that shows the size of the force (by the length of the arrow) and the direction in which the force is acting. Adding the forces (to determine the net force) requires that students take the directions of each force into account. Mathematically, this requires students to add or subtract vectors, which is different than adding or subtracting numbers.

After developing an understanding of forces, students learn to apply these ideas in increasingly more difficult contexts. A context would include systems with multiple objects that interact, including the analysis of collisions, and systems that rotate. The analysis of collisions requires an understanding of Newton's third law, which lies at the foundation of the principle of the conservation of momentum. Objects that rotate provide a context for students to revisit kinematics and dynamics using angular quantities, such as angular velocity, angular acceleration, and torque. In addition, many physics courses contain a separate unit on gravity. This unit typically includes the introduction to the universal law of gravity and the study of planetary motions using Kepler's laws.

Students need these multiple contexts to gain a deeper understanding of the fundamental principles of force and motion. Many physics teachers encourage or require students to pursue research projects or to conduct independent experiments. One such area that is rich in possibilities is the study of simple machines. Studying simple machines gives students practice in identifying forces and analyzing motions in situations that include levers, pulleys, and gears; they also explore the conditions required for static equilibrium (balancing).

It should be noted that although the sequence of ideas we describe is very common to most high school physics courses, variability does exist. One variation is to introduce the concept of energy prior to the concept of force. Another variation is the "physics first" program, with physics being taught in ninth grade, followed by chemistry and biology. If this sequence is adopted, it is generally recommended that the physics course focus more on concept development and experimental methods than on traditional problem solving that requires a more sophisticated understanding of mathematics.

Formative Assessment Reminder

Now that you have the background on this new series and this book, let's not forget the formative purpose of these probes. Remember—a probe is not formative unless you use the information from the probe to modify, adapt, or change your instruction so that students have increased opportunities to learn the important ideas related to force and motion. As a companion to this book, NSTA has co-published the book *Science Formative Assessment: 75 Practical Strategies for Linking Assessment, Instruction, and Learning* (Keeley 2008). In this book you will find strategies to use with the probes to

Preface

facilitate elicitation of student thinking, support metacognition, spark inquiry, encourage discussion, monitor progress toward conceptual change, encourage feedback, and promote self-assessment and reflection. We hope these probes and the techniques you can use along with them will stimulate new ways of assessing your students, create conducive environments for learning, promote richer discussions, and help you discover and use new knowledge about teaching and learning.

References

American Association for the Advancement of Science (AAAS). 1993. *Benchmarks for science literacy*. New York: Oxford University Press.

Bransford, J., A. Brown, and R. Cocking. 2000. *How people learn: Brain, mind, experience, and school*. Washington, DC: National Academies Press.

Driver, R., A. Squires, P. Rushworth, and V. Wood-Robinson. 1994. *Making sense of secondary science: Research into children's ideas*. London: Routledge-Falmer.

Keeley, P. 2005. *Science curriculum topic study: Bridging the gap between standards and practice*. Thousand Oaks, CA: Corwin Press and Arlington, VA: NSTA Press.

Keeley, P. 2008. *Science formative assessment: 75 practical strategies for linking assessment, instruction, and learning*. Thousand Oaks, CA: Corwin Press and Arlington, VA: NSTA Press.

Mundry, S., P. Keeley, and C. J. Landel. 2009. *A leader's guide to science curriculum topic study*. Thousand Oaks, CA: Corwin Press.

National Research Council (NRC). 1996. *National science education standards*. Washington, DC: National Academies Press.

Robertson, W. 2002. *Force and motion: Stop faking it! Finally understanding science so you can teach it*. Arlington, VA: NSTA Press.

Thornton, R. and D. Sokoloff. 1998. Assessing student learning of Newton's laws: The force and motion conceptual evaluation. *American Journal of Physics* 66 (4): 228–351.

Appendix

The book *Science Curriculum Topic Study: Bridging the Gap Between Standards and Practice* (Keeley 2005) describes the process for creating a standards-based and research-informed assessment probe. The book is made up of 147 single-page curriculum topic study (CTS) guides intended for science educators to use to

- learn more about a science topic's content,
- examine instructional implications,
- identify specific learning goals and scientific ideas,
- examine the research on student learning,
- consider connections to other topics,
- examine the coherency of ideas that build over time, and
- link understandings to state and district standards.

The CTS guides use national standards (*Benchmarks for Science Literacy* [AAAS 1993] and the *National Science Education Standards* [NRC 1996]) and research (Driver et al. 1994 and others) in a systematic study process that deepens teachers' understanding of particular science topics. For example, Figure 1 (p. xx) shows a CTS guide that was used to study the topic of *gravitational force*. Then, as shown in Figure 2 (p. xxi), the resources in that guide were used to create a probe on *gravity*.

In Figure 2, the ideas from the standards in the left-hand column in bold type were matched with commonly held student ideas as cited in the research. These ideas are in the right-hand column, also in bold type. Then, based on the infor-

Preface

Figure 1. Curriculum Topic Study (CTS) Guide: Gravitational Force

THE CURRICULUM TOPIC STUDY GUIDES **215**

Standards- and Research-Based Study of a Curricular Topic

GRAVITATIONAL FORCE

Section and Outcome	**Selected Sources and Readings for Study and Reflection** Read and examine *related parts* of:
I. Identify Adult Content Knowledge	**IA:** ***Science for All Americans*** ▸ Chapter 4, *Forces of Nature,* pages 55–57 **IB:** ***Science Matters: Achieving Scientific Literacy*** ▸ Chapter 1, *Gravity,* pages 9–14 ▸ Chapter 9, *Quantum Gravity,* pages 132–133 ▸ Chapter 12, *General Relativity,* pages 167–171
II. Consider Instructional Implications	**IIA:** ***Benchmarks for Science Literacy*** ▸ 4G, *Forces of Nature* general essay, page 93; grade span essays, pages 94–96 **IIB:** ***National Science Education Standards*** ▸ Grades K–4, Standard B essay, pages 123, 126 ▸ Grades 5–8, Standard B essay, page 149, 154 ▸ Grades 9–12, Standard B essay, pages 177–178
III. Identify Concepts and Specific Ideas	**IIIA:** ***Benchmarks for Science Literacy*** ▸ 4G, *Forces of Nature,* pages 94–97 **IIIB:** ***National Science Education Standards*** ▸ Grades K–4, Standard B, *Position and Motion of Objects,* page 127 ▸ Grades 5–8, Standard B, *Motion and Forces,* page 154 ▸ Grades 9–12, Standard B, *Motions and Forces,* pages 179–180
IV. Examine Research on Student Learning	**IVA:** ***Benchmarks for Science Literacy*** ▸ 4G, *Forces of Nature,* page 340 **IVB:** ***Making Sense of Secondary Science: Research Into Children's Ideas*** ▸ Chapter 16, *Magnetism and Gravity,* page 126 ▸ Chapter 23, *Gravity,* pages 163–167
V. Examine Coherency and Articulation	**V:** ***Atlas of Science Literacy*** ▸ *Gravity,* pages 42–43 ▸ *Laws of Motion,* pages 62–63
VI. Clarify State Standards and District Curriculum	**VIA:** ***State Standards:*** Link Sections I–V to learning goals and information from your state standards or frameworks that are informed by the results of the topic study. **VIB:** ***District Curriculum Guide:*** Link Sections I–V to learning goals and information from your district curriculum guide that are informed by the results of the topic study.

Visit www.curriculumtopicstudy.org for updates or supplementary readings, Web sites, and videos.

Source: Keeley, P. 2005. *Science curriculum topic study: Bridging the gap between standards and practice.* Thousand Oaks, CA: Corwin Press and Arlington, VA: NSTA Press.

mation in this chart, the probe "Gravity Rocks!" on page 171, for example, was developed to see whether students recognize that gravitational force decreases with distance from the Earth's surface and whether students confuse gravitational potential energy with gravitational force. This process was generally used throughout this volume, with additional research derived from physics education research.

Science Curriculum Topic Study was developed as a professional development resource for teachers with funding from the National Science Foundation's Teacher Professional Continuum Program. The book is accompanied by *A Leader's Guide to Science Curriculum Topic Study* (Mundry, Keeley, and Landel 2009). The leader's guide includes a workshop model and a CD-ROM of resources, templates, tools, and PowerPoint slides for leading a professional development session on how to create a probe. After using the probes in the present book, you might like to try to create your own probe using the curriculum topic study (CTS) guides/topics on at the beginning of each section in this book (pp. 12, 68, and 140).

Preface

Figure 2. Curriculum Topic Study (CTS) Chart for Developing a Probe on Gravity*

CTS Section III (K–12 Concepts and Specific Ideas)	CTS Section IV (K–12 Research on Learning)
Gravity • Things near the earth fall to the ground unless something holds them up. BSL K–2 • The earth's gravity pulls any object on or near the earth toward it without touching it. BSL 3–5 • **Every object exerts gravitational force on every other object. The force depends on how much mass the objects have and on how far apart they are.** The force is hard to detect unless at least one of the objects has a lot of mass. BSL 6–8 • The sun's gravitational pull holds the earth and other planets in their orbits, just as the planets' gravitational pull keeps their moons in orbit around them. BSL 6–8 • Gravity is the force that keeps planets in orbit around the sun and governs the rest of the motion in the solar system. Gravity alone holds us to the earth's surface and explains the phenomena of the tides. NSES 5–8 • **Gravitational force is an attraction between masses. The strength of the force is proportional to the masses and weakens rapidly with increasing distance between them.** BSL 9–12 • **Gravitation is a universal force that each mass exerts on any other mass. The strength of the gravitational attractive force between two masses is proportional to the masses and inversely proportional to the square of the distance between them.** NSES 9–12	**Gravity** • Elementary students typically do not see gravity as a force. They see the phenomenon of falling as 'natural' with no need to attribute it to force. BSL • **Some high school students believe gravity increases with height above earth's surface** or are not sure whether the force of gravity would be greater on a wooden ball or lead ball of the same size. BSL • High school students have difficulty conceptualizing gravitational forces as interactions and that the magnitudes of the gravitational forces that two objects of different mass exert on each other are equal. BSL • Children's ideas about Earth's gravity depend on their conception of a spherical earth and how 'down' is interpreted. MSSS • Holding, rather than pulling, seems to be a common perception of gravity bound up with the idea of gravity being associated with air pushing down and an atmosphere of air that prevents things from floating away. MSSS • Only some objects exert gravitational force and gravity only affects heavy things. MSSS • Earth's magnetism and spin are associated with gravity. MSSS • **Some students confuse gravity with potential energy in assuming a higher force of gravity at higher heights.** MSSS • …There is no force of gravity in water which is why things float, there is less gravity in water, there is gravity in water but it acts upward, or gravity only acts on the parts of the body above the surface of the water. MSSS • Some students think gravity only applies to objects on earth. MSSS • Gravity as 'molecules of gravity' in air. MSSS

*The ideas from the standards in the left-hand column in bold type were matched with commonly held student ideas as cited in the research. These ideas are in the right-hand column, also in bold type.

BSL= *Benchmarks for Science Literacy* (AAAS 1993)
NSES= *National Science Education Standards* (NRC 1996)
MSSS= *Making Sense of Secondary Science* (Driver et al. 1994)

Acknowledgments

We would like to thank the teachers and science coordinators we have worked with for their willingness to field-test probes, share student data, and contribute ideas for additional assessment probe development. We would also like to thank our colleagues at the Maine Mathematics and Science Alliance (MMSA) (*www.mmsa.org*) and the Blake School in Minneapolis who support us in this work. In addition, we thank the Math-Science Partnership directors, our professional development colleagues, and our university partners throughout the United States whom we have had the pleasure of sharing the *Uncovering Student Ideas in Science* work with in various projects.

We would also like to thank our reviewers for providing useful feedback to improve the original manuscript of this book: Marilyn Decker, former director of science with the Boston Public Schools; Dr. Arthur Eisenkraft, University of Massachusetts; Dr. Pam Kraus, FACET Innovations; Dr. Stamatis Vokos, Seattle Pacific University; Dr. Gerald Wheeler, former executive director of the National Science Teachers Association; and John Whitsett, coordinator of curriculum and instruction for the Fond du Lac (Wisconsin) School District and former physics teacher. We especially thank Dr. Jim Minstrell at FACET Innovations for taking the time to write a foreword for this first volume in the new series of *Uncovering Student Ideas* books. Jim has certainly set the gold standard for us in terms of what it really means to attend to student thinking when planning for instruction and teaching for understanding.

About the Authors

Page Keeley is the senior science program director at the Maine Mathematics and Science Alliance (MMSA) where she has worked since 1996. She directs projects in the areas of leadership, professional development, linking standards and research on learning, formative assessment, and mentoring and coaching, and she consults with school districts and organizations nationally. She was the principal investigator on three National Science Foundation grants: the Northern New England Co-Mentoring Network; Curriculum Topic Study: A Systematic Approach to Utilizing National Standards and Cognitive Research; and PRISMS: Phenomena and Representations for Instruction of Science in Middle School. She is the author of 10 books (including this one): four books in the *Curriculum Topic Study* series (Corwin Press); four volumes in the *Uncovering Student Ideas in Science: 25 Formative Assessment Probes* series (NSTA Press); and *Science Formative Assessment: 75 Practical Strategies for Linking Assessment, Instruction, and Learning* (Corwin Press and NSTA Press).

Most recently she has been consulting with school districts, Math-Science Partnership projects, and organizations throughout the United States on building teachers' capacity to use diagnostic and formative assessment. She is frequently invited to speak at national conferences, including the annual conference of the National Science Teachers Association. She led the People to People Citizen Ambassador Program's Science Education delegation to South Africa in 2009 and to China in 2010.

Page taught middle and high school science for 15 years; in her classroom she used formative assessment strategies and probes long before there was a name attached to them. Many of the strategies in her books come from her experiences as a science teacher. During her time as a classroom teacher, Page was an active teacher leader at the state and national level. She received the Presidential Award for Excellence in Secondary Science Teaching in 1992 and a Milken National Distinguished Educator Award in 1993. She was the AT&T Maine Governor's Fellow for Technology in 1994, has served as an adjunct instructor at the University of Maine, is a Cohort 1 Fellow in the National Academy for Science and Mathematics Education Leadership, and serves on several national advisory boards.

Prior to teaching, she was a research assistant in immunology at the Jackson Laboratory of Mammalian Genetics in Bar Harbor, Maine. She received her BS in life sciences from the University of New Hampshire and her MEd in secondary science education from the University of Maine. Page was elected the 63rd president of the National Science Teachers Association for the 2008–2009 term. In 2009 she received the National Staff Development Council's Susan Loucks-Horsley Award for her contributions to science education leadership and professional development.

About the Authors

Dr. Rand Harrington is the preK–12 science department chair and science curriculum coordinator for The Blake School in Minneapolis. He began his teaching career in 1980 as a middle school science teacher in California after receiving a degree in environmental science at Western Washington University. In 1985, after teaching and traveling throughout the world, he returned to school and received a second bachelor's degree in physics and then completed both his master's degree and PhD in physics at the University of Washington.

As a science teacher, Rand had long been interested in understanding how people learn, and he soon joined the Physics Education Research Group at the University of Washington under the leadership of Lillian McDermott. While working with this group, he taught and helped develop curriculum materials for *Physics by Inquiry*, a curriculum for preservice teachers, as well as *Tutorials in Introductory Physics*, which is used in many introductory physics courses. Rand was able to pursue his own interests in electricity and magnetism and eventually wrote his PhD thesis on identifying and addressing the difficulties students have with understanding electric phenomena.

After graduation from the University of Washington, Rand accepted an assistant professor appointment at the University of Maine, where he founded the Physics Education Research Group (originally called LRPE) and collaborated with the Maine Mathematics and Science Alliance (MMSA). In 1998 he was awarded a Higher Education SEED Foundation grant from MMSA and the Maine Department of Education to work with preservice teachers and to reform the introductory physics courses for nonscience majors at the University of Maine. In addition he received a National Science Foundation grant to examine best practices in science teaching. He has served on the ETS Physics SAT II test construction committee and on the American Association of Physics Teachers committee on research in physics education.

In 1999, he left Maine to help start a "Physics First" high school science program at the Harker School in San Jose, California. During that time, he adapted materials for a high school curriculum based on modeling, *Tutorials in Introductory Physics*, and *Physics by Inquiry*. He also became interested in computer-based tutorials and the effectiveness of online homework such as WebAssign and Mastering Physics. In 2005, Rand assumed his present position at The Blake School in Minneapolis.

He has served as a consultant for several independent schools, is a reviewer for the *American Journal of Physics*, teaches a summer course for undergraduate science and engineering majors at the University of Minnesota, received the Juliet Nelson Award for Excellence in Teaching, and has taught physics to Tibetan monks as part of the Science for Monks program in Dharamsala, India. His most recent interests are finding effective methods to "extend the thinking" of students at all grade levels and to use the computer as a tool for effective learning.

Introduction

Force and Motion: Research, Teaching, and Student Ideas

> We are convinced that the more probes that teachers use, the sounder their appreciation of their students' understanding, the more interesting they and their students will find their teaching, and the better will be the learning that follows.
>
> —Richard White and Richard Gunstone
> *Probing Understanding*

There is little doubt that the topics of force and motion present difficult challenges for both students and their teachers. Students have had more direct, personal experience with the ideas of force and motion than they have had with perhaps any other topics in the science curriculum; thus, they often come to class with fully formed and strongly held beliefs. Not all of these beliefs are consistent with a scientific view, however. When teachers take deliberate actions to understand what students believe, and why, they are taking an important first step toward improving their teaching practices.

Historical Background

Swiss child psychologist Jean Piaget identified some of these difficulties with the concepts of force and motion in interviews with children over a half century ago (Inhelder and Piaget 1958). During interviews on a task identified as "Conservation of Motion in a Horizontal Plane," he found that "the subject provides contradictory explanations: Light balls go farther because they are easier to set in motion. Larger ones go father because they are stronger. There is an absence of laws." In the late 1950s and early 1960s, curriculum develop-

Introduction

ers and researchers such as Robert Karplus, Arnold Arons, and others made use of the work of Piaget and developed science curricula using various conceptual change models borrowed from cognitive psychologists. Even as curricula have evolved over the decades since the 1960s, however, student difficulties persist and teachers who listen carefully will still hear the echoes of Piaget's young research subjects (Arons 1977; Karplus and Thier 1967).

In addition to numerous mathematical difficulties that students experience in describing motion (a field of study called kinematics), deep conceptual difficulties also exist. Extensive interviews with students and experiences in the classroom have helped teachers and researchers develop a better understanding of these conceptual difficulties. The results from this work have been used to create various diagnostic tests, primarily for secondary students and undergraduates, including the commonly used Force Concept Inventory (Hestenes, Wells, and Swackhamer 1992) and the Mechanics Baseline Test (Hestenes and Wells 1992) as well as parts of the physics Diagnoser project (FACET Innovations 2008) and the Force and Motion Conceptual Evaluation (Thornton and Sokoloff 1998).

Why Students Have Difficulties With Force and Motion Concepts

What is the source of these difficulties and why are they so persistent? One area that seems to underlie many of the problems is related to a person's ability to differentiate between a quantity and a change in that quantity. For example, understanding the difference between height (a quantity) and rate of growth (a change in that quantity) is similar to understanding the difference between position (a quantity) and speed (a change in that quantity). Perhaps the most difficult ratio for students to understand in the study of motion is the ratio we call acceleration (the change in velocity in one unit of time). Acceleration involves a rate of change of a quantity that is itself a rate (i.e., the rate of a rate). The fact that a child's growth rate can be decreasing while the child continues to grow taller is as difficult for students to grasp as the idea that acceleration can be decreasing even as the object continues to speed up (Arons 1983, 1984a, 1984b).

In addition to the difficulties students have with ratios and proportional reasoning, many students hold deeply ingrained beliefs about the nature of force. Their difficulties may be semantic and inadvertently reinforced by commonly used phrases such as, "I will force you to…," "may the force be with you," or "the force of gravity." The use of the word *force* in these statements implies that force is an object or a property of that object rather than *an interaction between objects* (the scientific definition). Teachers can help even our youngest students by making sure they use the word *force* as a description of an interaction, rather than as a property of a single object (Touger 1991).

Many students are also deeply committed to the idea that motion does not happen without a cause. However, in nature, it is only the change in motion that demands a causal interaction and not motion itself. This is highly counterintuitive because in our everyday experiences we rarely experience the absence of interactions (forces). The change in motion of an object is a result of **unbalanced** forces, and constant motion is the result of **balanced** forces or the absence of interactions altogether (no forces acting). In the case of constant motion, one of the forces is often "hidden" in the interaction that we call friction. This misleads us to think that the active force (such as a push from the hand acting on the object) is the only force acting. If an object is sliding on a surface, when the hand stops pushing, then the object slows down and stops. This slowing is a result of the friction interaction by the surface act-

Introduction

ing on the object. However, students rarely, if ever, experience what happens when the hand is removed and *there are no other forces acting.* In this case, the object would continue moving in a straight line, neither speeding up nor slowing down indefinitely (or until the object interacts with another object).

One of our goals should be to move students from what is called Aristotelian thinking (motion implies force) to Newtonian thinking (change in motion implies an unbalanced force). Researchers have found that less than 25% of our high school or college physics students typically come to our classrooms with a Newtonian view of forces (Hake 1998). After instruction, this number typically does not exceed 50% even though many of these same students can excel in applying the mathematics necessary to solve traditional problems. This evidence suggests that it is important to help even our youngest students develop an understanding of force as an *interaction* between two objects—in place of their *mis*understanding that it is a property of, or an action caused by, a single object.

Common Instructional Difficulties Related to Teaching About Gravitational Interactions

The subtleties related to understanding gravitational interactions also can present challenges to both teachers and their students. These difficulties can be categorized into three major areas: (1) understanding the nature or cause of gravity, (2) understanding the difference between weight and mass, and (3) understanding the motion of falling objects.

The Nature or Cause of Gravity

One of the best examples of how to elicit student ideas related to the cause of gravity comes from the research of a former high school physics teacher and current senior research scientist and co-founder of FACET Innovations, Jim Minstrell (Minstrell and Kraus 2005). Minstrel showed his students a small weight hanging from a spring scale. He then placed the scale with the weight inside a bell jar. A bell jar is a large glass dome that is sealed at the bottom. A pump is then used to remove the air from the inside of the jar. Minstrell asked his students to predict what would happen to the scale reading as air was removed from inside the jar. The results were striking: A large number of students believed that the scale reading either would be reduced or would go to zero as the air was removed! Children's books, such as one of the books in the popular Magic School Bus series (Cole and Degen 1990) show astronauts who appear "weightless" when they leave their space ship. These images inadvertently reinforce incorrect ideas about the relationship between gravity and atmosphere. It also does not help when terms such as *weightless* and *zero gravity* are used in common (nonscientific) speech. It can be difficult to help students understand that "gravity" is the name we give to a universal interaction between any two masses and that this interaction happens "at a distance"—meaning the objects do not have to touch.

Understanding the Difference Between Weight and Mass

Perhaps the most difficult (and contentious) idea for students to understand is the difference between *weight* and *mass*. To further complicate matters, even scientists do not always agree on how to define these words. For example, in Europe, scientists define *weight* as the force needed to hold you up (such as the reading on a bathroom scale). Using this definition, an object that is not being held up (e.g., when an object is in free fall) would be called weightless. Astronauts circling the Earth would be described as weightless using this definition.

Introduction

However, in the United States, scientists define *weight* as the gravitational force. A bathroom scale can indicate your weight but only if the scale itself is not accelerating. For example, if you weigh yourself on a scale inside an elevator that is speeding up (while the elevator is going upward), the scale would read more than your actual weight, but less than your actual weight when the elevator slows down. In the United States, this reading would be called your "apparent weight." Defining weight as the gravitational force also means that a person is never "weightless" because gravity is a long-range force that does not require contact. Astronauts in the Space Shuttle may feel weightless, but there is still gravitational force acting on them by the Earth. This gravitational force (called their "weight" by those in the United States) is why the astronauts move in a circular orbit around the Earth. You can see why these ideas can easily confuse students!

To avoid these difficulties, we can introduce students to a different concept called mass, which is a property of matter. (Don't worry, this definition is agreed upon by all scientists!) This property is the same regardless if you are in an elevator, on the Moon, or in orbit.

One way to help students understand the difference between these concepts is to use a spring scale to find the weight of an object and to use a balance to find the mass of an object. A spring scale will have a smaller reading on the Moon than on the Earth (the object will weigh less on the Moon). However, if an object can balance 10 g of mass on the Earth, then it will also balance the same 10 g on the Moon.

Scientists try to keep these ideas separate by using different units for mass and weight. In the metric system, mass is measured in grams or kilograms and weight is measured in newtons. In the English system of measurement, weight is measured in pounds and mass is measured in "slugs." To further complicate matters, the word *slug* is so uncommon that food packagers prefer to mix the units, often equating weight units (like pounds or ounces) to mass units (like grams or kilograms). If you look at a scale in the produce section of a supermarket, you will see that both pounds and ounces *and* grams and kilograms are used. If your students struggle with these ideas, tell them they are not alone. Several years ago NASA lost a very expensive satellite that did not go where NASA wanted the satellite to go. After a lot of troubleshooting, NASA identified the problem: Software engineers had used units of pounds (the gravitational force acting on a mass of one slug) while scientists had used units of newtons (the gravitational force acting on a mass of one kilogram)! In other words, one NASA team had used metric units and another had used English units in their calculations and no one noticed until it was too late.

Understanding the Motion of Falling Objects

Galileo discovered a very important physical principle: Gravitational force is directly proportional to the mass of an object. This means that an object with twice the mass of another object will also weigh twice as much as that object. According to some accounts, Galileo discovered this by dropping cannonballs with different masses off the Tower of Pisa. He noticed that the more massive ball reached the ground at about the same time as a less massive ball; therefore, the force on the more massive ball must be more than the force on the less massive ball. (If the forces were the same, then the less massive ball would reach the ground first!)

Why is this idea difficult for our students to accept? Because when our students drop objects, the objects will likely NOT reach the ground at the same time—the more massive object will reach the ground first. Skydivers, skiers, and bikers are all aware of this fact—that is, the heavier you are, the faster you go. So

Introduction

who is correct? You can see why students start to think that the science we teach them is not only irrelevant, but also incorrect. The answer is that objects only fall with the same acceleration in the absence of air resistance or friction. Experienced physics teachers learn to minimize the effect of air resistance by using small heavy objects and dropping them only a short distance. It is not clear if this strategy is successful for all students because many will walk away wondering why experiments in science class conflict with their everyday experiences.

Stepping-Stones: Emerging Concepts and Language

There are many challenges to teaching science, but one of the most difficult choices we face as teachers is to know when to accept a child's emerging concept, called a stepping-stone, in place of a scientific idea that most scientists would agree is important and correct from a scientific point of view. Stepping-stone concepts are actually central to effective learning, yet they are not always what come to the minds of physicists when they are asked about important ideas in physics. In a recent symposium at the National Academies of Science, at which a reorganization of K–12 science education around core science ideas was being considered, it was proposed that candidates for core ideas include both stepping-stone ideas as well as scientific ideas (Wiser and Smith 2009).

When looked at through the eyes of more advanced students, some of young children's emerging ideas could be judged as being incorrect or downright wrong. For example, when is $F_{net} = ma$ a correct statement of Newton's law of motion? We know that this is true only when several assumptions are made, such as when applied to a system whose mass is not changing or a system that is not moving too fast (relative to the speed of light). However it would be misleading to claim that $F_{net} = ma$ is wrong. At some point we should teach our students that all of the relationships we use to describe nature are in fact models that provide only an approximation of what we can observe directly.

Teachers are often unsure about when to use the word *weight* and when to use the word *mass* with their students. For older students, this distinction is important and we often attempt to correct students who misuse these terms. However, the use of the word *weight* in younger grades should be considered a stepping-stone concept (matter has weight) preceding the core scientific idea that matter has mass. Younger students often mistake the word *mass* with the phonetically similar word *massive* and further confuse their understanding of mass with the concept of volume.

Other stepping-stone concepts are related to the use of the word *distance* in place of *position* or *displacement* and the use of the word *time* when we really mean *time interval*. Both of these concepts merge when we choose zero time and zero position as our starting points (as is common in math textbooks). Older students often confuse these concepts, so it is tempting to differentiate these ideas at a young age. However, it is our experience that it is best to wait to differentiate between these concepts until early middle school.

Several learning progressions are currently being investigated in the physical sciences. Learning progressions are defined as "descriptions of the successively more sophisticated ways of thinking about a topic that can follow one another as children learn about and investigate a topic over a broad span of time" (NRC 2007, p. 214). Because learning progressions are empirically tested, they will provide the much-needed research to help teachers decide when to use a stepping-stone in lieu of a core scientific idea.

In constructing the probes in this book, we have tried to follow this guide: We generally use the words *weight, distance,* and *time* in probes designed for younger students while

Introduction

we use *mass, position, displacement,* and *time interval* in probes for older students. However, we encourage teachers to use those terms that best match their own learning goals regardless of the wording that was chosen for any individual probe.

Implementing the Force and Motion Probes

The probes in this book can be used as windows into student thinking before, during, or after instruction. It should be noted that if a probe is used before instruction, teachers must carefully plan their curricula so that, after completing the probe, students have an opportunity to develop the relevant ideas. If a probe is administered before instruction, and there is no immediate follow-up, students may feel frustrated at not knowing the "correct answer" and teachers may feel compelled to provide students with direct answers without the necessary background or experiences. If the probes are used during instruction, the teacher must be sure to present other activities that reinforce the science concept on which the probe is based. If the probes are used after instruction, the teacher uses the information gained from the probes to plan for further learning opportunities that will help students who are still struggling with the concept. The advantage of using the probes after instruction—still formatively and not as summative assessments—is that students are better prepared to participate in classroom discussion. Students must feel comfortable sharing their ideas in the discussions, and the teacher must carefully manage the discussions so that all ideas are valued.

Classroom discussions are also wonderful opportunities to reinforce the necessary components of a scientific explanation when the teacher requires students to state their claims, the evidence in support of those claims, and the explanations that connect the claims to the evidence (Krajcik et al. 2006). If discussions are conducted during or after instruction, then students will have better access to direct evidence to support their claims.

Below are examples of how two probes in the book could be used as formative assessments both before and after instruction:

(Before Instruction)
Scenario 1: Fifth-Grade Unit on Motion Using the "Rolling Marbles" Probe (p. 59)

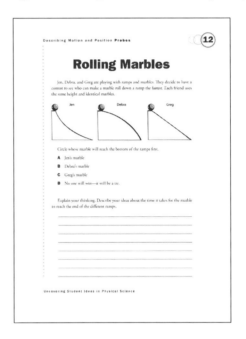

Teacher: Today we are going to be scientists! We are going to study the motion of a marble rolling down hills. Let us look at three different shapes of a hill. Which hill do you think the marble will get to the bottom of first? Draw a circle around that hill. Now let's share with each other.

Alisha: I think the hill that is a straight line will be the fastest. This is because the marble has the shortest distance to travel.

Michael: I don't know, but I think maybe the hill that is steep at the end. This is because the ball will be moving the fastest at the finish line.

Introduction

Georgia: I think it might be the hill that is steep at the beginning because the marble will start rolling really fast and get ahead of the other marbles.

Teacher: To be good scientists, we need to test our ideas, but we have to make sure our tests are fair. What do we need to do to make sure our tests are "fair"?

Jimmy: I think they must use the same size marble.

Lorenzo: And the marbles must start at the same height!

Mayumi: The hills should be made out of the same material.

Teacher: Here are three flexible tubes. Work with your partners and see if you can find out what shape track will help the marble get to the bottom first.

(After Instruction)
Scenario 2: Eighth-Grade Physical Science Class Using the "Finger Strength Contest" Probe (p. 127)

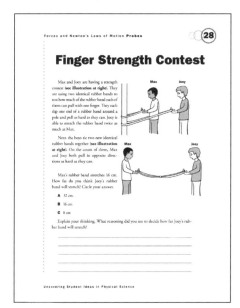

The students have just completed a unit on Newton's three laws of motion.

Teacher: What would the length of a stretched rubber band tell us about the forces acting on the rubber band?

Willie: If the rubber band is longer, I think that means there is a greater force.

Teacher: Can you give me an example to support this idea?

Shirley: Yes. If I pull harder, the rubber band gets longer ... just like this! [She demonstrates this to the class with a rubber band between her fingers.]

Teacher: Because of what Shirley and Willie described, a rubber band can be used as a "force meter."

Teacher hands out the probe called "Finger Strength Contest."

Teacher: Please answer the questions as you can on your own. Try to write an explanation that cites evidence to support your answer.

When the students have completed the probe, the teacher collects the results and quickly scans each paper. She keeps a tally to see how many students say the change in the length of the rubber bands will be the same and how many think the change in the length of each rubber band will be different. She notices that more than half the students believe the change in the lengths of the rubber bands will be different, a violation of Newton's third law.

Teacher: The results show us that about half of you believe that the rubber bands will have a different length and the other half believes the rubber bands will have the same length. I want you to work in groups of three or four to come up with an explanation for either of these answers and to find evidence to support that answer—but don't actually perform the

Introduction

experiment. On your individual whiteboards, please write your claim at the top, the evidence to support that claim, and an explanation (connect the evidence to your explanation). If you cannot agree as a group, then you may complete two different whiteboards.

After students complete the task, the teacher collects the whiteboards and displays them in front of the room. She arranges the whiteboards so that they are grouped according to the explanations and the claims.

Teacher: Let's look at the evidence that was used to support your claims. Do I have a volunteer from each group to describe the evidence that the groups used in their explanations?

Michelle: At first we thought the rubber bands would be different lengths because they pull with different forces. But then we remembered the lab we did last week where we attached two force probes together and pulled on them. No matter how we pulled, the forces were equal! So they must be the same.

Dillon: But this just doesn't make sense to us. Our evidence is that if you pull twice as much, the rubber band will stretch twice as much. So if one person is stronger, then that person's rubber band should also stretch twice as much!

Shemar: Yes, we thought the same thing. But now we agree with Michelle's group. This is also like the spring we put between two bathroom scales. No matter how much we pushed on one scale, the other scale would always have the same reading. So the spring is kind of like a force equalizer. Even if we try to pull differently, the forces always equal out.

A Final Caveat: Teachers Beware!

In choosing the probes in this book, we have attempted to find examples that connect basic principles to real-life experiences. However, there are a few probes that ask students to imagine an experience that they have never had—such as living in a world without friction or one in which there were no other forces acting on an object. In these cases, we have tried to be clear by not mixing real-life situations with these "ideal" frictionless or massless experiences found in a typical physics textbook. To avoid some difficulties, we recommend that you be ready to listen to all of your students' answers and their explanations before being too judgmental about what is right or wrong. Even in the Teacher Notes that follow each probe, we give a standard scientific response based on a few simple assumptions. There may be times when a student may not make the same assumptions, and although his or her answer may be different from ours, it could still be quite correct based on that student's own interpretation of the problem.

We leave you now with the words of the well-known cognitive psychologist David Ausubel: "The most important single factor influencing learning is what the learner knows. Ascertain this and teach accordingly" (Ausubel, Novak, and Hanesian 1978). While it is interesting in and of itself to discover what your students are really thinking, remember that assessment is not formative unless you use the information you have uncovered in the assessment to guide your instruction and promote learning.

References

Arons, A. 1977. *The various language: An inquiry approach to the physical sciences*. New York: Oxford University Press.

Arons, A. 1983. Student patterns of thinking and reasoning, I. *The Physics Teacher* (Dec.): 576–581.

Arons, A. 1984a. Student patterns of thinking and reasoning, II. *The Physics Teacher* (Jan.): 21–26.

Arons, A. 1984b. Student patterns of thinking and reasoning, III. *The Physics Teacher* (Feb.): 88–93.

Introduction

Ausubel, D., J. Novak, and H. Hanesian. 1978. *Educational psychology: A cognitive view.* 2nd ed. New York: Holt, Rinehart, and Winston.

Cole, J., and B. Degen. 1990. *The magic school bus lost in the solar system.* New York: Scholastic.

FACET Innovations. 2008. Diagnoser Project. *www.facetinnovations.com/daisy-public-website/fihome/resources/diagnoser*

Hake, R. 1998. Interactive-engagement versus traditional methods: A six-thousand-student survey of mechanics test data for introductory physics courses. *American Journal of Physics* 66: 64–74.

Hestenes, D., and M. Wells. 1992. A mechanics baseline test. *The Physics Teacher* 30: 159.

Hestenes, D., M. Wells, and G. Swackhamer. 1992. Force concept inventory. *The Physics Teacher* 30: 141.

Inhelder, B., and J. Piaget. 1958. *Growth of logical thinking: From childhood to adolescence.* New York: Basic Books.

Karplus, R., and H. Thier. 1967. *A new look at elementary school science.* Chicago: Rand McNally.

Krajcik, J., A. Novak, C. Gleason, and J. Mahoney. 2006. Creating a classroom culture of scientific practices. In *Exemplary science in grades 5-8: Standards-based success stories,* ed. R. Yager, 85–96. Arlington, VA: NSTA Press.

Minstrell, J., and P. Kraus. 2005. Guided inquiry in the science classroom. In *How students learn: History, mathematics, and science in the classroom,* ed. M. S. Donovan and J. D. Bransford. Washington, DC: National Academies Press.

National Research Council (NRC). 2007. *Taking science to school: Learning and teaching science in grades K–8.* Washington, DC: National Academies Press.

Thornton, R., and D. Sokoloff. 1998. Assessing student learning of Newton's laws: The force and motion conceptual evaluation. *American Journal of Physics* 66 (4): 228–351.

Touger, J. 1991. When words fail us. *The Physics Teacher* 29 (2): 90–95.

White, R., and R. Gunstone. 1991. *Probing understanding.* London: Falmer Press.

Wiser, M., and C. Smith. 2009. How does cognitive development inform the choice of core ideas in the physical sciences? Draft of commissioned paper for NRC Conference: Expert Meeting on Core Ideas in Science. Washington, DC: National Academies of Science. *www7.nationalacademies.org/bose/Core_Ideas_Background_Materials.html*

Section 1
Describing Motion and Position

	Concept Matrix	12
	Related Curriculum Topic Study Guides	13
	Related NSTA Press Books, NSTA Journal Articles, and NSTA Learning Center Resources	13
1	How Far Did It Go?	15
2	Skate Park	19
3	Following Jack: Part 1	23
4	Following Jack: Part 2	27
5	Go-Cart Test Run	31
6	Checking the Speedometer	35
7	Speed Units	39
8	Just Rolling Along	43
9	Crossing the Finish Line	47
10	NASCAR Racing	51
11	Roller Coaster Ride	55
12	Rolling Marbles	59
13	String Around the Earth	63

Concept Matrix
Probes #1–#13

PROBES	#1 How Far Did It Go?	#2 Skate Park	#3 Following Jack: Part 1	#4 Following Jack: Part 2	#5 Go-Cart Test Run	#6 Checking the Speedometer	#7 Speed Units	#8 Just Rolling Along	#9 Crossing the Finish Line	#10 NASCAR Racing	#11 Roller Coaster Ride	#12 Rolling Marbles	#13 String Around the Earth
GRADE LEVEL USE →	1–4	6–12	6–12	6–12	6–12	6–12	6–12	6–12	8–12	6–12	6–12	1–5	8–12
RELATED CONCEPTS ↓													
acceleration		X							X	X	X	X	
average speed							X		X			X	
changing speed		X											
circumference													X
clock readings						X	X						
constant speed		X	X	X	X				X				
displacement						X			X				
distance	X			X	X								
graph					X	X							
instantaneous speed									X				
measurement	X												
position	X		X	X	X	X			X				
proportion													X
ratio						X	X						X
speed		X	X	X	X	X	X	X	X	X	X	X	
time			X	X	X							X	
time intervals			X	X	X	X		X	X			X	
uniform motion		X	X	X	X				X				
units							X						
velocity											X	X	

Related Curriculum Topic Study Guides*

Describing Position and Motion
Forces
Graphs and Graphing
Motion
Observation, Measurement, and Tools

*Guides will be found in Keeley, P. 2005. *Science Curriculum Topic Study: Bridging the Gap Between Standards and Practice.* Thousand Oaks, CA: Corwin Press and Arlington, VA: NSTA Press. Each Curriculum Topic Study Guide shows the reader how to Identify Adult Content Knowledge, Consider Instructional Implications, Identify Concepts and Specific Ideas, Examine Research on Student Learning, Examine Coherency and Articulation, and Clarify State Standards and District Curriculum.

Related NSTA Press Books, NSTA Journal Articles, and NSTA Learning Center Resources

NSTA Press Books

American Association for the Advancement of Science (AAAS). 2001. *Atlas of science literacy.* Vol. 1. (See "Laws of Motion" map, pp. 62–63 and "Ratios and Proportionality" map, pp. 118–119.) Washington, DC: AAAS.

Eichinger, J. 2009. *Activities linking science with math, 5–8.* Arlington, VA: NSTA Press.

Horton, M. 2009. *Take-home physics: High impact, low-cost labs.* Arlington, VA: NSTA Press.

Keeley, P. 2005. *Science curriculum topic study: Bridging the gap between standards and practice.* Thousand Oaks, CA: Corwin Press and Arlington, VA: NSTA Press.

Konicek-Moran, R. 2008. Bocce anyone? In *Everyday science mysteries: Stories for inquiry-based science teaching*, 13–146. Arlington, VA: NSTA Press.

Morgan, E., and K. Ansberry. 2007. Roller Coasters. In *More Picture-Perfect Science Lessons: Using Children's Books to Guide Inquiry, K–4*, 133–141. Arlington, VA: NSTA Press.

Robertson, W. 2002. *Force and motion: Stop faking it! Finally understanding science so you can teach it.* Arlington, VA: NSTA Press.

NSTA Journal Articles

Abisdris, G., and A. Phaneuf. 2007. Using a digital video camera to study motion. *The Science Teacher* (Dec.): 44–47.

Ashbrook, P. 2008. The early years: Roll with it. *Science and Children* (Summer): 16–18.

Fechheim, J., and J. Nelson. 2007. Science sampler: Walk this way. *Science Scope* (Feb.): 50–52.

King, K. 2005. Making sense of motion. *Science Scope* (Feb.): 22–26.

Marshall, J., B. Horton, and J. Austin-Wade. 2007. Giving meaning to the numbers. *The Science Teacher* (Feb.): 36–41.

NSTA Learning Center Resources

NSTA Podcasts:
http://learningcenter.nsta.org/products/podcasts.aspx?lid=hp
Speed and Velocity
Inertia and Acceleration

NSTA SciGuides:
http://learningcenter.nsta.org/products/sciguides.aspx?lid=hp
Force and Motion

NSTA SciPacks:
http://learningcenter.nsta.org/products/scipacks.aspx?lid=hp
Force and Motion

NSTA Science Objects:
http://learningcenter.nsta.org/products/science_objects.aspx?lid=hp
Force and Motion
Force and Motion: Position and Motion

Describing Motion and Position

How Far Did It Go?

Before the car moves

After the car moves and stops

Gracie wants to measure the distance that her toy car travels. She places her car next to a measuring tape as shown in the first picture. She pushes the car. The second picture shows how far Gracie's car traveled until it stopped. Gracie measures the distance her car moved.

Circle the number of measurement units that best describes how far Gracie's car moved.

A 2

B 4

C 6

D 8

E 10

Describe how you figured out your answer.

___
___
___
___
___

Describing Motion and Position

How Far Did It Go?

Teacher Notes

Before the car moves

After the car moves and stops

Purpose
The purpose of this assessment probe is to see whether students recognize that units of distance traveled must be measured with a measurement device from the starting point to the ending point. The probe reveals whether students take into account the nonzero origin on a measurement scale or if they merely read off the number on the end point.

Related Concepts
distance, measurement, position

Explanation
The best answer is C: 6. The toy car's starting point is positioned at the 2-unit mark—that is, the car did not begin at the zero mark on the measurement scale. In this probe, distance is measured from the back of the car at the starting point of 2 units to the back of the car at the ending point (8-unit mark). The distance traveled equaled 6 units, between the 2-unit mark and the 8-unit mark.

Administering the Probe
This probe can be used with elementary students who are learning and practicing linear measurement skills. The teacher can model the probe by using a measuring tape or by drawing a picture for the students; a class discussion can follow.

Related Ideas in *National Science Education Standards* (NRC 1996)

K–4 Position and Motion of Objects
★ An object's motion can be described by tracing and measuring its position over time.

Related Ideas in *Benchmarks for Science Literacy* (AAAS 1993, 2009)

K–2 Manipulation and Observation
★ Measure length in whole units of objects using rulers and tape measures.

★ Indicates a strong match between the ideas elicited by the probe and a national standard's learning goal.

Describing Motion and Position

Related Research

- Mathematical studies reveal that few children recognize that any point on a measurement scale can serve as a starting point. They tend to read off whatever number is the end point. Even students up through fifth grade have been shown to have this tendency (Lindquist and Kouba 1989).
- When students are asked to measure the length of an object, they often begin their measurement on 1 rather than at the 0 point (Rose, Minton, and Arline 2007).
- One of the largest gaps in mathematics performance between minority students and Caucasian students is in the area of measurement (Lubienski 2003).

Suggestions for Instruction and Assessment

- Give students opportunities to measure the same length or distance from different starting points so that they realize that the units of length or distance are the same, regardless of different ending points when the starting point changes.
- Encourage students to focus on the lengths of the units rather than only on the numbers on the measurement scale.
- Encourage students to differentiate between *where* an object ends up (i.e., is now) from *how far* that object has gone.
- Explicitly point out to students who are just learning to use rulers and other length devices where the zero starting point is. Also, help them understand what the other markings on a length device, such as a ruler, represent.
- Consider modifying the probe so that the starting point is at the 1-unit mark. Because the research indicates that some students start their measurements on the 1-mark, rather than the 0-mark, a modified version of this probe may indicate whether students are using this intuitive, though incorrect, measurement rule.
- By the end of fifth grade, teachers should have given students many opportunities to use different measurement tools, techniques, and units for measuring distance and time.

References

American Association for the Advancement of Science (AAAS). 1993. *Benchmarks for science literacy.* New York: Oxford University Press.

American Association for the Advancement of Science (AAAS). 2009. Benchmarks for science literacy online. *www.project2061.org/publications/bsl/online*

Keeley, P. 2008. *Science formative assessment: 75 practical strategies for linking assessment, instruction, and learning.* Thousand Oaks, CA: Corwin Press and Arlington, VA: NSTA Press.

Lindquist, M., and V. Kouba. 1989. Measurement. In *Results from the Fourth Mathematics Assessment of the National Assessment of Educational Progress,* ed. M. Lindquist, 35–43. Reston, VA: National Council of Teachers of Mathematics.

Lubienski, S. 2003. Is our teaching measuring up? Race-, SES-, and gender-related gaps in measurement achievement. In *Learning and teaching measurement: 2003 yearbook,* ed. D. H. Clements and G. Bright, 282–292. Reston, VA: National Council of Teachers of Mathematics.

National Research Council (NRC). 1996. *National science education standards.* Washington, DC: National Academies Press.

Rose, C., L. Minton, and C. Arline. 2007. *Uncovering student thinking in mathematics.* Thousand Oaks, CA: Corwin Press.

Describing Motion and Position

Skate Park

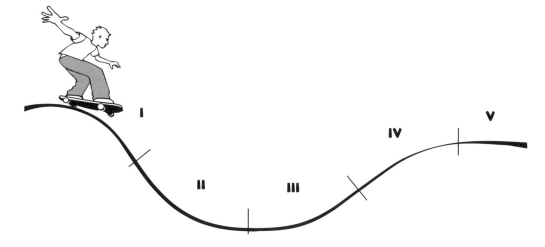

Billy is riding his skateboard in a skate park and comes to a steep hill. He coasts down the hill and then coasts back up the other side as shown in the picture above. Circle the phrase that best describes Billy's motion in each of the labeled (I, II, III, IV, V) sections above.

I   speeding up   slowing down   neither speeding up nor slowing down

II   speeding up   slowing down   neither speeding up nor slowing down

III   speeding up   slowing down   neither speeding up nor slowing down

IV   speeding up   slowing down   neither speeding up nor slowing down

V   speeding up   slowing down   neither speeding up nor slowing down

Explain your thinking about how to best describe Billy's motion at the different sections of the hill.

_____
_____
_____
_____
_____

Uncovering Student Ideas in Physical Science

Skate Park

Teacher Notes

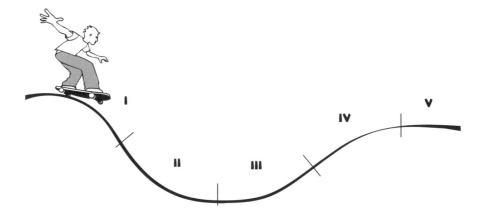

Purpose
The purpose of this assessment probe is to elicit students' descriptions of the motion of an object. The probe uses the context of rolling down and up a hill to determine if students recognize when an object speeds up or slows down or does neither (maintains a constant speed).

Related Concepts
acceleration, changing speed, constant speed, speed, uniform motion

Explanation
The best answers are I—speeding up; II—speeding up; III—slowing down; IV—slowing down; and V—neither slowing down nor speeding up (constant speed). As Billy goes down the hill on the skateboard, he is speeding up. As he coasts back up the hill, he will be slowing down. On the level section, he will eventually slow and come to a stop if there is friction between the wheels of the skateboard and the ground. If friction is negligible, then the skateboard will coast without speeding up or slowing down.

This motion can also be explained using the concept of energy. As the skateboarder loses gravitational potential energy (coming down the hill), he gains kinetic energy (he speeds up). As he goes back up the hill, the kinetic energy is transferred to gravitational potential energy and the skateboarder slows down. On the flat section, the gravitational potential energy stays the same, so the skateboarder neither speeds up nor slows down (as long as the friction force is negligible).

Administering the Probe
This probe is best used with middle and high school students. It is designed so that students will use everyday language like *speed up, slow down*, or *neither speed up nor slow down* before using technical terms. It can also be used with elementary students at an observational level by eliminating section V on the illustration, unless elementary students have the tools and ability to

Describing Motion and Position

measure uniform motion. If students have previously encountered the term *constant speed* and are familiar with its use, you may substitute it for *neither speed up nor slow down*. If students predict that the skateboarder will slow down in section V of the illustration, then teachers should probe for whether they correctly recognize the existence of a frictional force or if they believe that the motion just "naturally runs out."

Related Ideas in *National Science Education Standards* (NRC 1996)

5–8 Motions and Forces
★ The motion of an object can be described by its position, direction of motion, and speed.

Related Ideas in *Benchmarks for Science Literacy* (AAAS 1993, 2009)

6–8 Motion
- An unbalanced force acting on an object changes its speed or direction of motion or both.

9–12 Motion
- Any object maintains a constant speed and direction of motion unless an unbalanced, outside force acts on it.

Related Research
- Naturally, children's ideas and descriptions of motion tend to be less differentiated than those of a physicist's. Children tend to see objects either at rest or moving; they infrequently focus on the period of change. They use everyday terms such as *going faster* in ambiguous ways, sometimes referring to the magnitude of the speed of an object and at other times referring to the speed increasing with time. For instance, a child might say that a car is speeding if it is moving very fast even if the car is not speeding up (Driver et al. 1994, p. 155).
- Younger children typically start to describe motion by identifying the direction in which an object moves, without regard to the speed of the object. As the sophistication of their ideas progress, they may use a "snapshot" description, in which they compare the speed of an object at different locations or instants. (In a "snapshot" description, students describe what is essentially a still photograph of an object, without looking at changes.) Eventually, older children can be led to describe how the speed of an object is changing at a specific location or instant (Dykstra and Sweet 2009).

Suggestions for Instruction and Assessment
- Students should have multiple opportunities to describe motion in different contexts.
- In middle school, students should move from qualitative to quantitative descriptions of moving objects (NRC 1996). Encourage students to come up with methods to measure whether an object's speed is increasing, decreasing, or staying the same.
- Motion detectors connected to computers—or microcomputer-based laboratories (MBLs)—have been found to be effective in helping students connect real-world motions to graphical representations of motion. Students can speed up or slow down as they walk in front of the detector while a graph is displayed in real time on a computer screen.

References
American Association for the Advancement of Science (AAAS). 1993. *Benchmarks for science literacy.* New York: Oxford University Press.

American Association for the Advancement of Science (AAAS). 2009. Benchmarks for science literacy online. *www.project2061.org/publications/bsl/online*

★ Indicates a strong match between the ideas elicited by the probe and a national standard's learning goal.

Driver, R., A. Squires, P. Rushworth, and V. Wood-Robinson. 1994. *Making sense of secondary science: Research into children's ideas.* London: RoutledgeFalmer.

Dykstra, D., and D. Sweet. 2009. Conceptual development about motion and force in elementary and middle school students, *American Journal of Physics* 77 (5): 468–476.

National Research Council (NRC). 1996. *National science education standards.* Washington, DC: National Academies Press.

Describing Motion and Position

Following Jack: Part 1

Josey and her little brother Jack are walking side by side, eating ice cream cones. Josey stops to talk to a friend. While she is talking, Jack's ice cream cone starts to drip at a steady rate as Jack walks away. When Josey finishes talking to her friend and realizes that Jack is no longer next to her, she looks down and notices these drops of ice cream on the ground from Jack's ice cream cone:

●●● ●   ●   ●   ●   ●   ●   ●   ●

Josey needs help figuring out how Jack was moving (walking) while she was talking. If Josey follows the drips, what can they tell her about Jack's motion? Circle the answer that best shows how Jack moved (walked) while Josey stopped to talk to her friend.

A The drips show that Jack started walking really slowly and then went faster and faster.

B The drips show Jack started out walking really fast and then slowed down and went slower and slower.

C The drips show that Jack started out walking slowly, then walked faster and continued to walk at that same speed.

D The drips show that Jack started out walking fast, slowed down, and then continued to walk at that same, steady speed.

Explain your thinking. Provide an explanation for your answer.

Uncovering Student Ideas in Physical Science   23

Describing Motion and Position

Following Jack: Part 1

Teacher Notes

Purpose
The purpose of this assessment probe is to identify how students interpret a motion diagram and whether they have an operational understanding of the concept of speed. The probe is designed to show whether students can interpret the intervals (i.e., the spaces between the dots) and distinguish between slow versus fast speed and increasing speed versus constant speed using a motion diagram.

Related Concepts
constant speed, distance, position, time, time intervals, speed, uniform motion

Explanation
The best answer is C: The drips show that Jack started out walking slowly, then walked faster and continued to walk at that same speed. Because each pair of dots—that is, a set of two dots; the distance between each two dots shows how far Jack traveled in one unit of time—represents the same time interval, Jack did not travel far when the dots are closer together. This means he started by moving very slowly. As the dots get farther apart, Jack was increasing his speed. When the dots become equally spaced apart, Jack was traveling *the same distance in the same amount of time*. This means he started to move at a constant and unchanging speed.

Administering the Probe
This probe can be used before or after middle and high school students have been introduced to ticker tape–types of representations. Although ice cream may not normally drip at a steady, constant rate, for the purpose of this probe, make sure students recognize that the drip time intervals are the same. You may want to use an arrow pointing to the increasingly spaced dots to indicate the direction in which Josey's brother is walking.

Describing Motion and Position

Related Ideas in *National Science Education Standards* (NRC 1996)

K–4 Position and Motion of Objects
- An object's motion can be described by tracing and measuring its position over time.

5–8 Motions and Forces
★ The motion of an object can be described by its position, direction of motion, and speed.

Related Ideas in *Benchmarks for Science Literacy* (AAAS 1993, 2009)

K–2 Motion
- Things move in many different ways, such as straight, zigzag, round and round, back and forth, and fast and slowly.

3–5 Models
★ Diagrams, sketches, and stories can be used to represent objects, events, and processes in the real world.

6–8 Communication
- Students should understand writing that incorporates circle charts, bar and line graphs, two-way data tables, diagrams, and symbols.

6–8 Displacing the Earth From the Center of the Universe
★ The motion of an object is always judged with respect to some other object or point.

Related Research
- Naturally, children's ideas and descriptions of motion tend to be less differentiated than those of a physicist. They tend to see objects either at rest or moving. The period of change is less frequently focused on by children. They use everyday terms such as *going faster* in ambiguous ways, sometimes referring to the magnitude of the speed of an object and at other times referring to the speed increasing with time (Driver et al. 1994, p. 155).
- A similar task, called the "Ticker-Tape Puzzle," was used in the 1970s to identify student levels of reasoning from concrete to formal operational (Fuller, Karplus, and Lawson 1977).

Suggestions for Instruction and Assessment
- To create their own motion graphs, students, while walking, can drop beans, small rocks, or other objects in equal time intervals as they speed up or slow down.
- Motion diagrams can be generated with time-lapsed strobe photographs (also available on the internet) or by using a ticker tape, a timer, and a long strip of paper.
- Provide students with different ticker tape–type motion diagrams and ask them to interpret the motions.
- Use the probe "Following Jack: Part 2," page 27, to determine whether students can translate a ticker tape representation of motion into a graph.

References
American Association for the Advancement of Science (AAAS). 1993. *Benchmarks for science literacy.* New York: Oxford University Press.

American Association for the Advancement of Science (AAAS). 2009. Benchmarks for science literacy online. *www.project2061.org/publications/bsl/online*

Driver, R., A. Squires, P. Rushworth, and V. Wood-Robinson. 1994. *Making sense of secondary science: Research into children's ideas.* London: RoutledgeFalmer.

★ Indicates a strong match between the ideas elicited by the probe and a national standard's learning goal.

Fuller, R., R. Karplus, and A. Lawson. 1977. Can physics develop reasoning? *Physics Today* (Feb.): 23–28.

National Research Council (NRC). 1996. *National science education standards.* Washington, DC: National Academies Press.

Describing Motion and Position

Following Jack: Part 2

Josey and her little brother Jack are walking side by side, eating ice cream cones. Josey stops to talk to a friend. While she is talking, Jack's ice cream cone starts to drip at a steady rate as Jack walks away. When Josey finishes talking to her friend and realizes that Jack is no longer next to her, she looks down and notices these drops of ice cream on the ground from Jack's ice cream cone:

• • • •   •   •   •   •   •   •   •   •

Josey needs help figuring out what Jack was doing. Which of the following position versus time graphs best shows how Jack moved (was walking) while he was eating his ice cream cone? Circle the letter of the best graph.

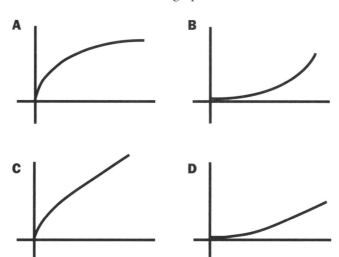

Explain your thinking. Describe how the graph you chose best matches Jack's motion.

Describing Motion and Position

Following Jack: Part 2

Teacher Notes

Purpose
The purpose of this assessment probe is to determine whether students can translate a motion diagram into a graph—a position versus time graph—to represent the motion of a moving object.

Related Concepts
constant speed, distance, graph, position, time, time intervals, speed, uniform motion

Explanation
The best answer is graph D. The steepness of the curve on a position versus time graph indicates the speed of an object. Graph D starts out almost flat (slow), as indicated by the closely spaced dots in the illustration on page 27. It then slightly curves to show how Jack speeds up as the dots spread out more within each interval. Finally, the graph becomes a straight line showing that Jack's speed eventually does not change (does not speed up or slow down).

Administering the Probe
This probe is best used with middle and high school students. It should be given after students have had an opportunity to qualitatively describe the representation on probe #3. Note that in physics, these graphs are typically referred to as position versus time graphs. However, in most middle school curriculum materials, these graphs are commonly referred to as distance versus time graphs. See the section on stepping-stone concepts in this book's introduction, page 5, for a discussion of *distance* and *position*.

Related Ideas in *National Science Education Standards* (NRC 1996)

K–4 Position and Motion of Objects
- An object's motion can be described by tracing and measuring its position over time.

Describing Motion and Position

5–8 Abilities of Inquiry
- ★ Use appropriate tools and techniques (including mathematics) to gather, analyze, and interpret data.
- ★ Use mathematics on all aspects of scientific inquiry.

5–8 Motions and Forces
- ★ The motion of an object can be described by its position, direction of motion, and speed.

9–12 Abilities of Inquiry
- Use technology and mathematics to improve investigations and communications.

Related Ideas in *Benchmarks for Science Literacy* (AAAS 1993, 2009)

3–5 Models
- Diagrams, sketches, and stories can be used to represent objects, events, and processes in the real world.

3–5 Constancy and Change
- ★ Often the best way to tell which kinds of change are happening is to make a table or graph of measurements.

6–8 Symbolic Relationships
- ★ Graphs can show a variety of possible relationships between two variables.

9–12 Symbolic Relationships
- ★ Tables, graphs, and symbols are alternative ways of representing data and relationships that can be translated from one to another.

Related Research

- A study by Boulanger (1976) found that training in proportional reasoning resulted in improved differentiation of speed, distance, and time among a group of 74 nine-year-old students (Driver et al. 1994).
- Students often experience difficulty interpreting the slope of a graph and sometimes confuse the height of the graph with the slope. Many students interpret graphs as literal pictures rather than symbolic representations (McDermott, Rosenquist, and van Zee 1987).
- Many students interpret distance/time graphs as the paths of actual journeys (Kerslake 1981). In addition, students confuse the slope of a graph and the graph's maximum or minimum value and do not know that the slope of a graph is a measure of rate (McDermott, Rosenquist, and van Zee 1987; Clement 1989).

Suggestions for Instruction and Assessment

- Before using this probe, use "Following Jack: Part 1" (pp. 23–26) to determine how students read a motion diagram.
- Reverse the task and challenge students to draw a motion diagram, using interval dots, for each of the four graphs.
- Students need tools to describe motion appropriately, including relevant vocabulary, graphical representations, and numerical formulas (Driver et al. 1994).

References

American Association for the Advancement of Science (AAAS). 1993. *Benchmarks for science literacy.* New York: Oxford University Press.

American Association for the Advancement of Science (AAAS). 2009. Benchmarks for science literacy online. *www.project2061.org/publications/bsl/online*

Boulanger, F. 1976. The effects of training in the proportional reasoning associated with the concept of speed. *Journal of Research in Science Teaching* 13 (2): 145–154.

Clement, J. 1989. The concept of variation and misconceptions in Cartesian graphing. *Focus on Learning Problems in Mathematics* 11 (1–2): 77–87.

★ Indicates a strong match between the ideas elicited by the probe and a national standard's learning goal.

Driver, R., A. Squires, P. Rushworth, and V. Wood-Robinson. 1994. *Making sense of secondary science: Research into children's ideas.* London: RoutledgeFalmer.

Kerslake, D. 1981. Graphs. In *Children's understanding of mathematics, 11–16,* ed. K. M. Hart, 120–136. London: John Murray.

McDermott, L. C., M. Rosenquist, and E. van Zee, 1987. Student difficulties in connecting graphs and physics: Examples from kinematics. *American Journal of Physics* 55 (6): 503–513.

National Research Council (NRC). 1996. *National science education standards.* Washington, DC: National Academies Press.

Describing Motion and Position

Go-Cart Test Run

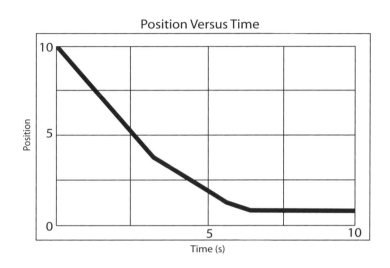

Jim and Karen have built a go-cart. They take their go-cart for a test run and graph its motion. Their graph is shown above. They show the graph to their friends. This is what their friends say:

Bill: "Wow, that was a steep hill! You must have been going very fast at the bottom."

Patti: "I think you were going fast at first, but then you slowed down at the end."

Kari: "I think you must have hit something along the way and come to a full stop."

Mort: "It looks like you were going downhill and then the road flattened out."

Circle the name of the friend you think best describes the motion of the go-cart, based on the graph. Explain why you agree with that friend.

_____

_____

_____

_____

_____

Go-Cart Test Run

Teacher Notes

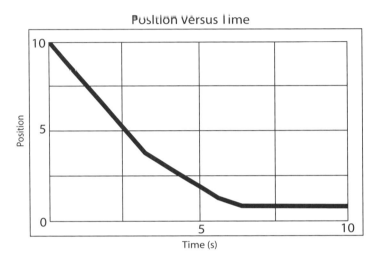

Purpose

The purpose of this probe is to examine how students interpret a graphical representation of motion. The probe is designed to reveal whether students interpret a motion graph *pictorially* or *mathematically*.

Related Concepts

clock readings, constant speed, graph, position, time, time intervals, speed, uniform motion

Explanation

Kari has the best answer: "I think you must have hit something along the way and come to a full stop." The speed of the go-cart is related to the steepness of the line on the graph. Between 0 and 3 seconds, the go-cart is moving at a constant speed. Between 3 and 6 seconds, the go-cart is still moving at a constant speed, but slower than the previous interval. Between 6 and 7 seconds, the go-cart is moving at a slower steady speed. From 7 seconds on, there is no change in position. The go-cart has stopped.

A graph can also be interpreted by the slope of the lines. A straight line with a steep slope indicates constant motion at a greater speed than a straight line with a more gradual slope. When the line on a graph is horizontal, that means that as time goes by, the position of an object is not changing. The horizontal line on this graph means that the go-cart stops at the end of the motion.

Administering the Probe

This probe is appropriate for middle school students and high school students. You may adapt the probe by eliminating the answer choices and having students "tell the story" indicated by the graph. For students who may be unfamiliar with position versus time graphs, it may be helpful to describe the y axis as the milepost or distance marker.

Describing Motion and Position

Related Ideas in *National Science Education Standards* (NRC 1996)

5–8 Abilities of Inquiry
★ Use appropriate tools and techniques (including mathematics) to gather, analyze, and interpret data.
★ Use mathematics on all aspects of scientific inquiry.

5–8 Motions and Forces
★ The motion of an object can be described by its position, direction of motion, and speed.

9–12 Abilities of Inquiry
• Use technology and mathematics to improve investigations and communications.

Related Ideas in *Benchmarks for Science Literacy* (AAAS 1993, 2009)

3–5 Constancy and Change
★ Often the best way to tell which kinds of change are happening is to make a table or graph of measurements.

6–8 Symbolic Relationships
★ Graphs can show a variety of possible relationships between two variables.

9–12 Symbolic Relationships
★ Tables, graphs, and symbols are alternative ways of representing data and relationships that can be translated from one to another.

Related Research
• Students of all ages often interpret graphs of situations as literal pictures rather than as symbolic representations of the situations (Leinhardt, Zaslavsky, and Stein 1990; McDermott, Rosenquist, and van Zee 1987).
• Many students interpret distance/time graphs as the paths of actual journeys (AAAS 1993, p. 351). (*Note:* This study refers to distance/time graphs, which is the more common way of labeling graphs with elementary and middle school students, in the same way that this probe uses position/time.)
• A study conducted by Kozhevnikov (2007) showed a correlation between spatial reasoning ability and the ability of students to correctly interpret motion graphs.

Suggestions for Instruction and Assessment
• Provide multiple opportunities for students to construct and interpret graphs so that you can see what students understand or misunderstand about graphs and graphing.
• Provide students with different types of motion graphs and have them make up stories about what the graphs show. Encourage discussions about the accuracy of their stories—that is, do the stories accurately reflect the information on the graphs? This strategy, popular in mathematics classes, can help students overcome the tendency to view graphs as a literal picture.
• Use motion detectors and students' real movements to help students construct a visual and kinesthetic understanding of position versus time. MBLs (microcomputer-based laboratories) are known to improve the development of students' abilities to interpret graphs (AAAS 2001). MBLs are particularly effective in helping middle school students understand that a graph is not a literal picture (Mokros and Tinker 1987).
• Most middle school science curricula use distance versus time graphs instead of posi-

---

★ Indicates a strong match between the ideas elicited by the probe and a national standard's learning goal.

tion versus time graphs. Although both types of graphs can be interpreted in the same way, teachers should help students understand the distinction between distance and position. In some textbooks, *distance* may mean *distance traveled* where *position* refers to the *location of an object*. In some special cases (such as when the motion starts at zero position) these terms mean the same thing. However, to avoid confusion, physics teachers tend to use *position,* which has a well-defined meaning.

References

American Association for the Advancement of Science (AAAS). 1993. *Benchmarks for science literacy.* New York: Oxford University Press.

American Association for the Advancement of Science (AAAS). 2001. *Atlas of science literacy.* Vol. 1. (See "Graphic Representation" map, pp. 114–115.) Washington, DC: AAAS.

American Association for the Advancement of Science (AAAS). 2009. Benchmarks for science literacy online. *www.project2061.org/publications/bsl/online*

Leinhardt, G., O. Zaslavsky, and M. Stein. 1990. Functions, graphs, and graphing: Tasks, learning, and teaching. *Review of Educational Research* (60): 1–64.

Kozhevnikov, M. 2007. Spatial visualization in physics problem solving. *Cognitive Science* 31 (4): 549–579.

McDermott, L., M. Rosenquist, and E. van Zee. 1987. Student difficulties in connecting graphs and physics: Example from kinematics. *American Journal of Physics* (55): 503–513.

Mokros, J., and R. Tinker. 1987. The impact of microcomputer-based labs on children's ability to interpret graphs. *Journal of Research in Science Teaching* (24): 369–383.

National Research Council (NRC). 1996. *National science education standards.* Washington, DC: National Academies Press.

Describing Motion and Position

Checking the Speedometer

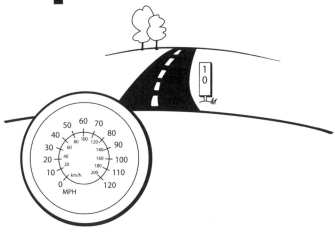

Maya's family bought a new car. Her mother asks Maya for help to make sure the speedometer is working. To check the speedometer, Maya's mother drives the car down a long, straight highway. On the side of the highway there is a marker for every mile. As Maya's mother drives at a constant speed, Maya writes down the time when they pass each of the numbered markers. Maya records the information in the data table shown below.

Place an X next to all of the ways that Maya could use the information in the data table to find the speed of the car:

____ 10/2:00   ____ 4/10

____ 2:04/14   ____ 11/1

____ 14/2:04   ____ 1/11

____ 1/1   ____ 4/4

____ 4/14   ____ 2/12

____ None of these ways would work!

Mile Marker	Time (hours : minutes)
10 mi	2:00
11 mi	2:01
12 mi	2:02
13 mi	2:03
14 mi	2:04

Describe your thinking about how you can find the speed of the car.

Uncovering Student Ideas in Physical Science

Describing Motion and Position

Checking the Speedometer

Teacher Notes

Purpose
One of the primary difficulties that students experience in determining the speed of a moving object is differentiating between quantities and changes in quantities. The purpose of this probe is to examine students' use of ratios that express displacement over a time interval to determine speed rather than using position and clock readings.

Related Concepts
clock readings, displacement, position, ratio, speed, time intervals

Explanation
There are two ways to find the speed of the car: 1/1 and 4/4. Speed is defined as the distance an object travels in one unit of time. Both of these ratios produce the same answer: a speed of 1 mile for each minute. This could also be described as a speed of 1 mile for every 1 minute, or 4 miles for every 4 minutes. To obtain this answer, students need to think about how far the car traveled in a given unit of time by subtracting two clock readings and then finding the difference between the two corresponding positions. This involves reasoning about displacement and a time interval (changes in quantities), rather than using position and clock readings (quantities). Some students may not recognize that miles for each minute are units for speed and may be more familiar with units of miles for each hour (commonly represented as *mph* or *miles per hour*).

Administering the Probe
This probe is appropriate for middle and high school students. Make sure students know that the / symbol means the same as the ÷ symbol. Explain that each marker has a number that shows the number of miles from the first marker. Each mile marker is 1 mile from the previous marker. This probe can also be used as a card sort, a formative assessment classroom technique (FACT) used to promote small-group discussion. Place each of the choices

Describing Motion and Position

on page 35 on separate cards and encourage students to sort the cards into those that represent the speed of the car and those that do not, explaining their reasoning for each choice (Keeley 2008).

Related Ideas in *National Science Education Standards* (NRC 1996)

5–8 Abilities of Inquiry
★ Use mathematics on all aspects of scientific inquiry.

5–8 Motions and Forces
★ The motion of an object can be described by its position, direction of motion, and speed.

9–12 Abilities of Inquiry
• Use technology and mathematics to improve investigations and communications.

Related Ideas in *Benchmarks for Science Literacy* (AAAS 1993, 2009)

6–8 Symbolic Relationships
★ Rates of change can be computed from differences in magnitudes and vice versa.

6–8 Numbers
★ Some interesting relationships between two variables include the variables always having the same difference or same ratio.

9–12 Computation and Estimation
★ Use ratios and proportions, including constant rates, in appropriate problems.

Related Research
• Early adolescents, as well as many adults, have difficulty with proportional reasoning. The difficulty is influenced by the format of the problem, the numbers used in the problem, the types of ratios used, and the context of the problem (AAAS 1993).
• Different ratio types produce different degrees of difficulty (AAAS 1993). For example, in studies by Heller et al. (1989), speed problems appear to be more difficult than exchange problems. Unfamiliarity with the problem situation causes even more difficulty when it occurs with a difficult ratio type (AAAS 1993, p. 360).
• Misailidou and Williams (2003) found little correlation between a child's computational ability and his or her ability to perform proportional reasoning. This result implies that students should be exposed to situations in which they are *required* to use proportions and are not allowed to avoid this type of reasoning by using algorithms to perform calculations.

Suggestions for Instruction and Assessment
• Although it may seem no more than a technical detail about mathematical notation, it is important for students to understand *a/b* as a representation of fractions, division, and ratios that are all equivalent and can be expressed as quotients (AAAS 2001).
• Some students may not recognize *mph* as a ratio of two numbers. It has been suggested (Arons 1977) that the word *per* be replaced with the words *for each* for greater clarity.
• An alternative approach to this problem would be to have students graph this information with position on the vertical axis and time (clock readings) on the horizontal axis. The speed of the car would be the slope of this line.
• Be aware that some textbooks contribute to the difficulty of differentiating between quantities and changes in quantities by their

---

★ Indicates a strong match between the ideas elicited by the probe and a national standard's learning goal.

careless use of the words *distance* and *time*. *Time* can mean either a clock reading (an instant in time) or a time interval. *Distance* is often used for either position or displacement, especially in middle school instructional materials. *Speed* is the ratio of displacement over a time interval but is usually represented mathematically in textbooks as *d/t*. It is common for students to interpret these quantities incorrectly as positions and clock readings. Some textbooks introduce the symbol Δ to indicate "a change in" as in Δ*x*/Δ*t*. It is not uncommon to also see Δ*d*, which could be interpreted as the change in the distance (a change in displacement would not make sense). The symbol *d* can also be ambiguous but usually means "distance traveled."

References

American Association for the Advancement of Science (AAAS). 1993. *Benchmarks for science literacy*. New York: Oxford University Press.

American Association for the Advancement of Science (AAAS). 2001. *Atlas of science literacy*. Vol. 1. (See "Ratios and Proportionality" research narrative and map, pp. 118–119.) Washington, DC: AAAS.

American Association for the Advancement of Science (AAAS). 2009. Benchmarks for science literacy online. *www.project2061.org/publications/bsl/online*

Arons, A. 1977. *The various language: An inquiry approach to physical science*. New York: Oxford University Press.

Heller, P., A. Ahlgren, T. Post, M. Behr, and R. Lesh. 1989. Proportional reasoning: The effect of two context variables, rate type, and problem setting. *Journal of Research in Science Teaching* (26): 205–220.

Keeley, P. 2008. *Science formative assessment: 75 practical strategies for linking assessment, instruction, and learning*. Thousand Oaks, CA: Corwin Press and Arlington, VA: NSTA Press.

Misailidou, C., and J. Williams. 2003. Diagnostic assessment of children's proportional reasoning. *Journal of Mathematical Behavior* 22: 335–368

National Research Council (NRC). 1996. *National science education standards*. Washington, DC: National Academies Press.

Describing Motion and Position

Speed Units

Max is thinking about different units that can be used to express the average speed of a car on a highway. Check off all the units that can be used to express the average speed of a car on a highway.

____ kilometers per hour   ____ inches per hour   ____ meters per minute

____ inches per second   ____ miles per minute   ____ centimeters per hour

____ feet per minute   ____ miles per hour   ____ kilometers per second

____ meters per hour   ____ feet per hour   ____ centimeters per second

Explain your thinking. How did you decide which units can be used to express the average speed of a car on a highway?

_____

_____

_____

_____

_____

_____

_____

Uncovering Student Ideas in Physical Science   39

Describing Motion and Position

Speed Units

Teacher Notes

Purpose
The purpose of this assessment probe is to determine whether students recognize that there is a variety of measurement units, described as ratios, that can be used to express average speed, even though some units may not be practical for the situation that exists.

Related Concepts
average speed, ratio, speed, units

Explanation
All 12 units in the list on page 39 can be used to express the average speed of the car. The question asks which units *can be* used, not which units are *best* used. Although miles per hour or kilometers per hour are the most familiar units used to measure and express the average speed of a car on a highway, it is still possible to express the speed by converting to other units (although some conversions may not be practical). For example, the average speed of a car traveling 50 miles per hour on the highway could be expressed as 73 feet per second. A car traveling at 80 kilometers per hour could also be expressed as traveling at 80,000 meters per hour, even though this conversion may not be practical unless the car was stuck in a traffic jam and barely moved over the course of an hour. Any average speed the car is traveling at can be converted to any of the units included on the list on page 39, even though the units may not be practical for expressing the average speed of the car.

Administering the Probe
This probe is best used with middle school and high school students. Make sure students understand that the intent of the probe is to check off *all* the units that are *possible* to use. Be careful not to cue them too much because you want to see if they select only units that are typically used for describing the speed of a car on a highway.

Describing Motion and Position

Related Ideas in *National Science Education Standards* (NRC 1996)

5–8 Abilities of Inquiry
- Use mathematics on all aspects of scientific inquiry.

5–8 Motions and Forces
- ★ The motion of an object can be described by its position, direction of motion, and speed.

9–12 Abilities of Inquiry
- Use technology and mathematics to improve investigations and communications.

Related Ideas in *Benchmarks for Science Literacy* (AAAS 1993, 2009)

6–8 Computation and Estimation
- ★ Determine what unit an answer should be expressed in from the units of the inputs to the calculation and be able to convert compound units (such as miles per hour to feet per second).

6–8 Communication Skills
- ★ Choose appropriate units for reporting various magnitudes.

9–12 Computation and Estimation
- ★ Use ratios and proportions, including constant rates, in appropriate problems.

Related Research
- Early adolescents, as well as many adults, have difficulty with proportional reasoning. The difficulty is influenced by the format of the problem, the numbers used in the problem, the types of ratios used, and the context of the problem (AAAS 1993).
- Different ratio types produce different degrees of difficulty (AAAS 1993). For example, in studies by Heller et al. (1989), speed problems appear to be more difficult than exchange problems. Unfamiliarity with the problem situation causes even more difficulty when it occurs with a difficult ratio type (AAAS 1993, p. 360).
- Some students may use the intuitive rule "more A, more B." If the speed is determined to be fast, then larger units of measurement are used (Stavy and Tirosch 2000).

Suggestions for Instruction and Assessment
- Use online conversion tables to show how units used to measure speed can be converted from one unit to another.
- As a follow-up to this probe, ask students to identify the measurement units that would be most practical to use to measure the average speed of a car as it is traveling on the highway.
- After students have identified the measurement units that would be most practical to use to measure the average speed of the car as it is traveling on the highway, challenge them to describe a scenario in which they might use the other units. For example, would there ever be an occasion to use inches per hour to describe the speed of a car? Is there a different object these units of speed (p. 39) would be more practical to use with in describing motion? When might you use the other units of speed with other objects or materials?

References
American Association for the Advancement of Science (AAAS). 1993. *Benchmarks for science literacy.* New York: Oxford University Press.

American Association for the Advancement of Science (AAAS). 2009. Benchmarks for science lit-

---

★ Indicates a strong match between the ideas elicited by the probe and a national standard's learning goal.

eracy online. *www.project2061.org/publications/bsl/online*

Heller, P., A. Ahlgren, T. Post, M. Behr, and R. Lesh. 1989. Proportional reasoning: The effect of two context variables, rate type, and problem setting. *Journal of Research in Science Teaching* (26): 205–220.

National Research Council (NRC). 1996. *National science education standards.* Washington, DC: National Academies Press.

Stavy, R., and D. Tirosch. 2000. *How students (mis-) understand science and mathematics: Intuitive rules.* New York: Teachers College Press.

Describing Motion and Position

Just Rolling Along

Jerome rolled a rubber ball across a very long table by giving the ball a very light push and then letting it roll across the table on its own. Six of his classmates observed the ball as it rolled.

Jerome wondered what happened to the speed of the ball after it left his hand. He asked the other students if they think it is possible to make the ball roll at a constant speed (*constant speed* means the ball is neither slowing down nor speeding up).

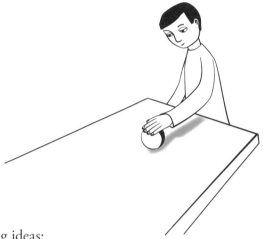

The students in Jerome's group shared the following ideas:

Anna: "It is not possible to make the ball roll at a constant speed."

Dev: "It is possible for the ball to roll at a constant speed if you tilt the table slightly downward."

Tad: "It is possible for the ball to roll at a constant speed if you tilt the table slightly upward."

Jack: "It is possible for the ball to roll at a constant speed if you make the table perfectly flat."

Byron: "It is possible for the ball to roll at a constant speed if you roll the ball really fast."

Talia: "It is possible for the ball to roll at a constant speed if you roll the ball really slow."

Circle the name of the student you think has the best idea. Explain why you think that is the best idea.

_____

_____

_____

_____

Uncovering Student Ideas in Physical Science

Describing Motion and Position

Just Rolling Along

Teacher Notes

Purpose

The purpose of this assessment probe is to elicit ideas about uniform motion. The probe is designed as a starting point to encourage students to use evidence and observations to support their ideas. The goal is for students to eventually develop and then test an operational understanding of the concept of speed.

Related Concepts

constant speed, displacement, speed, time intervals, uniform motion

Explanation

The best answer is Dev's: "It is possible for the ball to roll at a constant speed if you tilt the table slightly downward [at a very small angle])". When level, the ball slows down and when at a steep angle, the ball speeds up. One can then reason (and then verify experimentally) that there must be an angle at which the ball neither speeds up nor slows down.

The correct answer can also be found by analyzing forces. There is an angle at which the frictional force is balanced by the component of the gravitational force along the direction of the ball's motion. When the table is level, the ball will slow down because the friction force by the table is acting on the ball. This force acts opposite the direction of motion. In addition, when the table is level, there is no gravitational force acting in the direction of motion.

Administering the Probe

This probe is appropriate for middle school and high school students. Make sure that students, particularly middle school students, understand what the term *constant speed* means even though it is defined in the probe. Have props (rubber ball and long table) available to illustrate the context of the probe. This probe can be used with the P-E-O strategy: *predict*, *explain* the reason for your prediction, and *observe*; if students' observations do not

Describing Motion and Position

fit their predictions, students revise their predictions and explanations (Keeley 2008). To observe this motion, students will need to roll the ball slowly. If the table is tilted too much, then the ball will continuously speed up. With just a small tilt, they should be able to find the place where part of the gravitational force that pulls the ball down the incline is offset by the rolling friction acting on the ball by the table.

This probe can be answered and tested from a purely kinematics point of view, without requiring an explanation of forces on the part of the teacher or students. It can also be used as a probe in Section 2, "Forces and Newton's Laws," if you are interested in probing further for students' explanations of the forces involved.

Related Ideas in *National Science Education Standards* (NRC 1996)

5–8 Motions and Forces
★ The motion of an object can be described by its position, direction of motion, and speed.

Related Ideas in *Benchmarks for Science Literacy* (AAAS 1993, 2009)

6–8 Motion
- An unbalanced force acting on an object changes its speed or direction of motion, or both

9–12 Motion
★ Any object maintains a constant speed and direction of motion unless an unbalanced outside force acts on it.

Related Research
- Many researchers have found that substantial numbers of students are strongly committed to the idea that constant speed implies that a constant force is being applied to a moving object (Driver et al. 1994, p. 158).
- This probe addresses a particular "problematic facet" (Minstrell 1992) that objects—even objects rolling on horizontal surfaces—slow down because of gravity. Students do not see the need to have the force that is changing the motion be related to the direction of motion.

Suggestions for Instruction and Assessment
- This probe can be used at the start of a unit on kinematics—the branch of physics that deals with the motion of a body or system without reference to force and mass—as a way to elicit ideas that students have prior to instruction. If used in this way, it should be immediately followed up with a hands-on experiment in which students test their predictions and provide supporting evidence for their ideas.
- Students will develop a wide variety of reasons to support their prediction (even if it is not correct). Understanding that there must be a point where the ball neither speeds up nor slows down is similar to understanding a "point of inflection" in mathematics and can be quite difficult for some students. These students would benefit from taking measurements, such as comparing the time it takes the ball to move across the first half of the table with the time it takes the ball to travel across the second half of the table. They should then be led to adjust the tilt of the table until these two times are the same.
- It is important for teachers to listen carefully to how students use key words such as *force, momentum,* and *energy*. How students use these words can provide a window into student thinking about ideas not yet introduced in the unit. Rather

---

★ Indicates a strong match between the ideas elicited by the probe and a national standard's learning goal.

than correcting any inaccurate uses, ask students what they mean by these terms and then redirect them to use more direct descriptions of the motion (such as speeding up, slowing down, or moving at constant speed).

- Be aware that many students will try to bring the concept of force into their explanations and therefore may have a difficult time observing the motion without a bias.
- This probe can be used in postinstruction to see if students have developed an operational definition of speed and if they understand how to design an experiment to test an idea.
- Be careful when using this probe that you do not imply that objects—even objects rolling on horizontal surfaces—slow down because of gravity, As described by some of the "facets of student knowledge" (see Minstrell 1992), some students do not see the need to have the force that is changing the motion be related to the direction of the motion.

References

American Association for the Advancement of Science (AAAS). 1993. *Benchmarks for science literacy.* New York: Oxford University Press.

American Association for the Advancement of Science (AAAS). 2009. Benchmarks for science literacy online. *www.project2061.org/publications/bsl/online*

Driver, R., A. Squires, P. Rushworth, and V. Wood-Robinson. 1994. *Making sense of secondary science: Research into children's ideas.* London: RoutledgeFalmer.

Keeley, P. 2008. *Science formative assessment: 75 practical strategies for linking assessment, instruction, and learning.* Thousand Oaks, CA: Corwin Press and Arlington, VA: NSTA Press.

Minstrell, J. 1992. Facets of students' knowledge and relevant instruction. In *Research in physics learning: Theoretical issues and empirical studies*, ed. R. Duit, F. Goldberg, and H. Niedderer, 110–128. Proceedings of an International Workshop: Research in Physics Learning: Theoretical Issues and Empirical Studies. Kiel, Germany.

National Research Council (NRC). 1996. *National science education standards.* Washington, DC: National Academies Press.

Describing Motion and Position

Crossing the Finish Line

Frances and Greg decide to have a footrace. Frances gets off to a good start and gets ahead of Greg. Greg catches up to Frances right at the finish line so that they cross the finish line at the same time. Immediately after crossing the finish line, both are still running at their same pace and Greg passes Frances. Their friends argue about who ran faster right at the finish line. This is what they say:

Donica: "I think Frances was faster at the finish line."

Hannah: "I think Greg was going the fastest at the finish line."

Evan: "I think they were running at the same speed the instant they crossed the finish line."

Circle the name of the student you most agree with. Explain why you think that is the best answer.

_____
_____
_____
_____
_____
_____
_____

Uncovering Student Ideas in Physical Science   47

Describing Motion and Position

Crossing the Finish Line

Teacher Notes

Purpose
The purpose of this assessment probe is to elicit students' ideas about comparing motions. The probe is designed to see if students differentiate between the concept of position and the concept of speed. The probe will also help to see if students understand the difference between average speed and speed at an instant (often called "instantaneous speed").

Related Concepts
acceleration, average speed, instantaneous speed, position, speed, time intervals

Explanation
The best response is Hannah's: "I think Greg was going the fastest at the finish line." This is because Greg is passing Frances at the finish line. When one person is passing another, the two persons cannot be traveling at the same speed. Ideally, we would like students to be able to reason that two runners have the same speed if the distance between the runners is not getting larger or smaller. The two runners pass each other twice: once in the beginning of the motion and another time at the end of the motion. When they are side by side, they do not have the same speed. However, if they start at the same position, and at an equal time later they are again at the same position, their average speeds are the same for that time interval. The more mathematically inclined students will often calculate "speed" by saying that both runners go the same distance in the same amount of time. This number represents the average speed and not the instantaneous speed at the finish line.

Administering the Probe
This probe is appropriate for upper middle school and high school students. An alternative way to administer this probe is to set up the actual physical situation. Using balls, show students the motion several times (this takes some practice!) and then have students yell out when they think the balls have the same speed.

Describing Motion and Position

If balls are not available, then students can be asked the same questions based on a wide variety of representations: motion diagrams (stroboscopic photograph), a table of data with a graph of the motion, or a table of data based on a video of the motion.

Related Ideas in *National Science Education Standards* (NRC 1996)

5–8 Motions and Forces
★ The motion of an object can be described by its position, direction of motion, and speed.

9–12 Abilities Necessary to Do Inquiry
- Use technology and mathematics to improve investigations and communications.

Related Ideas in *Benchmarks for Science Literacy* (AAAS 1993, 2009)

6–8 Symbolic Relationships
- Rates of change can be computed from differences in magnitudes and vice versa.

9–12 Motion
- All motion is relative to whatever frame of reference is chosen, for there is no motionless frame from which to judge all motion.

Related Research
- This speed comparison task was the basis of Trowbridge's research (Trowbridge and McDermott 1981) into student understanding of speed. For their work, students rolled two steel balls down two metal U tracks (one level and one tilted at an angle). Students then observed the motion and indicated when they thought the two balls would have the same speed. It is often surprising how common it is for students to believe that the balls have the same speed when they pass each other.
- Students who say that Frances was faster at the finishing line may be using reasoning that fits the intuitive rule "more A, more B" (Stavy and Tirosch 2000). In this case, they see that Frances is ahead for most of the race so her speed must be the greatest at the finish line.

Suggestions for Instruction and Assessment
- After administering this probe, teachers should discuss and work through the problem. They should mention to students that there was an actual court case in which a judge confused speed with position and incorrectly found a driver guilty of speeding.
- Show students a position versus time graph of a similar motion. The graph provides additional evidence that two objects can be at the same location but not have the same speed. Point out on the graph that when the two lines intersect, they do not have the same slope.
- A similar question, the Jogger and Sprinter Elicitation, can be found on the Diagnoser website at *diagnoser.com*, under the Position and Distance Unit.

References
American Association for the Advancement of Science (AAAS). 1993. *Benchmarks for science literacy.* New York: Oxford University Press.

American Association for the Advancement of Science (AAAS). 2009. Benchmarks for science literacy online. *www.project2061.org/publications/bsl/online* .

National Research Council (NRC). 1996. *National science education standards.* Washington, DC: National Academies Press.

★ Indicates a strong match between the ideas elicited by the probe and a national standard's learning goal.

Stavy, R., and D. Tirosch. 2000. *How students (mis-) understand science and mathematics: Intuitive rules.* New York: Teachers College Press.

Trowbridge, D., and L. McDermott 1981. Investigation of student understanding of the concept of velocity in one dimension. *American Journal of Physics* 48 (12): 1020–1028.

Describing Motion and Position

NASCAR Racing

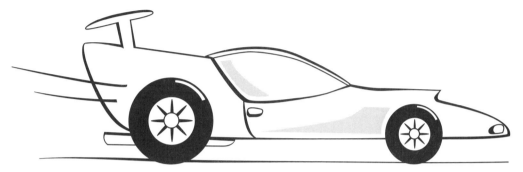

Kenisha and her friends are looking at a NASCAR article on the sports page of their local newspaper. Kenisha sees the word *velocity* mentioned several times. She wonders what the word *velocity* means. She asks her friends and this is what they say:

Silas: "I think it is the term used when something moves really fast."

Jade: "I think it is the scientific word for speed. *Speed* and *velocity* mean the same thing."

Ayla: "I think the words *velocity* and *acceleration* mean the same thing."

Omar: "I think it describes the speed and the direction in which something moves."

LaVonn: "I think it is the rate at which the speed of something is changing."

Terrell: "I think it is used to describe the average speed when something moves at different speeds."

Circle the name of the person you think has the best idea about what the word *velocity* means.

Explain why you agree with that person.

_____

_____

_____

_____

NASCAR Racing

Teacher Notes

Purpose
The purpose of this assessment probe is to determine what students mean when they use words to describe motion, such as *speed, velocity,* and *acceleration*. The probe is designed to reveal students' initial meaning of the word *velocity* before formally encountering the word in instruction.

Related Concepts
acceleration, speed, velocity

Explanation
Omar has the best idea about the meaning of velocity: "I think it describes the speed and the direction in which something moves." In everyday language, the terms *speed* and *velocity* are often used interchangeably. Physicists distinguish between the two terms. Velocity is speed in a given direction. Speed describes how fast something moves; velocity describes how fast something moves and in which direction it is moving. When velocity changes, either the speed, direction, or both are changing. *Constant velocity* means the speed and direction are unchanging. A NASCAR race car can have a constant speed around the track but not constant velocity because on a race track the car's direction changes when it moves along the curves.

Calculating the average speed of an object requires dividing the distance traveled by the time it takes to travel that distance. Calculating the average velocity requires dividing the displacement (a change in position) traveled by the time interval. The peculiar result of this calculation is that if you return to your original starting position, your average velocity will be zero, but your average speed will not be zero.

Administering the Probe
This probe can be used with middle school and high school students. Middle school is the time when the term *velocity* is introduced. Although the word *acceleration* is used in one of the distracters in the probe, middle school

Describing Motion and Position

students are not expected to understand acceleration quantitatively (although several states' science standards do include it). Even though national standards do not deal with acceleration in a quantitative way, the term is often qualitatively introduced in middle school curriculum materials and students hear the word in an everyday context.

Encourage students to share where they may have heard the word *velocity* before, even if they are not sure of its meaning. This probe can be combined with a formative assessment classroom technique called the Scientific Terminology Inventory Probe (STIP) (Keeley 2008). STIPs are short, simple questionnaires that teachers use formatively to ascertain students' familiarity with scientific terms they will encounter in an instructional unit (terms such as *speed, velocity,* and *acceleration*).

Related Ideas in *National Science Education Standards* (NRC 1996)

5–8 Abilities of Inquiry
- Use mathematics in all aspects of scientific inquiry.

5–8 Motions and Forces
★ The motion of an object can be described by its position, direction of motion, and speed.

9–12 Abilities of Inquiry
★ Use language appropriately [under Communicate and Defend a Scientific Argument].

Related Ideas in *Benchmarks for Science Literacy* (AAAS 1993, 2009)

9–12 Computation and Estimation
- Use ratios and proportions, including constant rates, in appropriate problems.

Related Research
- Researchers have pointed out that students need opportunities to develop the appropriate language tools to describe motion. These tools include vocabulary, graphical representations, and numerical formulas (Driver et al. 1994).
- Terms like *velocity* and *acceleration* are not commonly used by school-age children before the terms have been introduced in science lessons (Driver et al. 1994).

Suggestions for Instruction and Assessment
- Be aware that students often encounter words used in science outside of school and often develop meanings for these scientific words before formally encountering them in school science. Take time to find out what students mean by words such as *speed, velocity,* and *acceleration* in order to correct their preconceptions.
- Use the Activity Before Concept (ABC) approach. Providing an experience before giving students the word or concept helps them connect the word to its meaning in context. Give students opportunities to measure and describe speed and then add direction so that students can see how velocity differs from speed (Eisenkraft 2006).
- The concept of velocity is needed when there are forces acting that change the direction in which an object is moving—for instance, when a ball rolls uphill and then rolls downhill or when a car drives around a corner at constant speed. The mathematical method of keeping track of direction requires a representation called a "vector" and is best left for high school age students (NCTM 2000).

References
American Association for the Advancement of Science (AAAS). 1993. *Benchmarks for science literacy.* New York: Oxford University Press.

★ Indicates a strong match between the ideas elicited by the probe and a national standard's learning goal.

American Association for the Advancement of Science (AAAS). 2009. Benchmarks for science literacy online. *www.project2061.org/publications/bsl/online* .

Driver, R., A. Squires, P. Rushworth, and V. Wood-Robinson. 1994. *Making sense of secondary science: Research into children's ideas.* London: RoutledgeFalmer.

Eisenkraft, A. 2006. ABC: Activity before concept. Presented at the National Science Teachers Association National Conference, St. Louis.

Keeley, P. 2008. *Science formative assessment: 75 practical strategies for linking assessment, instruction, and learning.* Thousand Oaks, CA: Corwin Press and Arlington, VA: NSTA Press

National Council of Teachers of Mathematics (NCTM). 2000. *Principles and standards for school mathematics.* Reston, VA: NCTM.

National Research Council (NRC). 1996. *National science education standards.* Washington, DC: National Academies Press.

Describing Motion and Position

Roller Coaster Ride

Kimi is riding on a roller coaster with her friends. One of her friends exclaims, "Wow, that is a huge acceleration!" After the ride was over, Kimi wonders what the word *acceleration* means. Put an X in front of all the motions below that are an example of *acceleration*:

____ When something is moving really fast.

____ When something is moving really slowly.

____ When you turn a corner going really fast.

____ When you are going in one direction, turn around, and go in the other direction.

____ You are going slowly, but then speed up.

____ You are going fast, but then slow down.

Explain what the word *acceleration* means to you. Provide reasons for the motions you selected to describe acceleration.

_____
_____
_____
_____
_____
_____
_____
_____
_____

Describing Motion and Position

Roller Coaster Ride

Teacher Notes

Purpose
The purpose of this assessment probe is to elicit students' meaning of words used to describe motion, such as *speed, velocity,* and *acceleration*. The probe is designed to reveal students' initial qualitative meaning of the word *acceleration* before formally encountering the word in a scientific context.

Related Concepts
acceleration, speed, velocity

Explanation
All of the motions except for "moving really fast" or "moving really slowly" are examples of acceleration. An object can be moving very fast or very slow but not experience acceleration. This is because a fast (or slowly) moving object is not changing its velocity. For the velocity to change, the object must be speeding up, slowing down, or changing direction. For an object that is changing direction (such as a ball being thrown up into the air), the velocity of the object is zero at the turnaround point, but the acceleration is not zero because the velocity of the object is always changing.

Administering the Probe
This probe can be used at the middle school and high school levels. Although middle school is the time when the term *acceleration* is qualitatively introduced, middle school students are not expected to quantitatively understand acceleration. Encourage students to share where they may have heard the word *acceleration* before, even if they are not sure of its meaning. This probe can be combined with a formative assessment classroom technique called the Scientific Terminology Inventory Probe (STIP) (Keeley 2008). STIPs are short, simple questionnaires used for formative purposes that help teachers ascertain students' prior familiarity with scientific terms (such as *speed, velocity,* and *acceleration*) before they are encountered in an instructional unit.

Describing Motion and Position

Related Ideas in *National Science Education Standards* (NRC 1996)

5–8 Abilities of Inquiry
- Use mathematics on all aspects of scientific inquiry.

5–8 Motions and Forces
★ The motion of an object can be described by its position, direction of motion, and speed.

9–12 Abilities of Inquiry
★ Use language appropriately [under Communicate and Defend a Scientific Argument].

Related Ideas in *Benchmarks for Science Literacy* (AAAS 1993, 2009)

9–12 Computation and Estimation
- Use ratios and proportions, including constant rates, in appropriate problems.

Related Research
- Researchers have pointed out that students need opportunities to develop the appropriate language tools to describe motion. The tools include vocabulary, graphical representations, and numerical formulas (Driver et al. 1994).
- Terms like *velocity* and *acceleration* are not commonly used by school-age children prior to their introduction in science lessons (Driver et al. 1994).
- Students may use the intuitive rule "more A, more B" and, conversely, "less A, less B" to describe acceleration—that is, they might believe that the faster an object moves, the more it accelerates; the slower an object moves, the less it accelerates (Stavey and Tirosch 2000).
- Some students think acceleration is due to an increasing force (Twigger et al. 1994).
- It is quite common for students to confuse the concept of acceleration with the concept of velocity (Trowbridge and McDermott 1981).

Suggestions for Instruction and Assessment
- Be aware that students often give meaning to scientific words before formally encountering them in school science. Teachers who take the time to find out what students' existing meanings of words like *speed, velocity,* and *acceleration* are will avert several learning barriers.
- Use the Activity Before Concept (ABC) technique. Providing an experience before giving students the word or concept helps them connect the word to its meaning in context. Provide opportunities for students to describe how the speed of an object is changing, using the words *speeding up, slowing down,* or *changing direction* before introducing the word *acceleration* (Eisenkraft 2006).
- Be aware that young students may equate going fast with a large acceleration. Later, many students may associate acceleration with "speeding up" and not recognize that an object is also accelerating when it slows down or changes direction. These same students may use the word *decelerating* to mean slowing down. Provide opportunities for students to compare how a physicist uses the word *acceleration* with how it is used in common, everyday language.
- Be aware that students interpret negative acceleration with slowing down. The sign of the acceleration is related to the direction that the object would have to move in order to speed up. A negative acceleration does not necessarily mean the object is

---

★ Indicates a strong match between the ideas elicited by the probe and a national standard's learning goal.

slowing down. Therefore, defining deceleration as negative acceleration is incorrect.
- The concept of acceleration is needed when forces act to change the velocity of an object. (This is an example of Newton's second law of motion.)
- The mathematical method of keeping track of direction requires a representation called a "vector" and is best left for high school age students (NCTM 2000).

References

American Association for the Advancement of Science (AAAS). 1993. *Benchmarks for science literacy.* New York: Oxford University Press.

American Association for the Advancement of Science (AAAS). 2008. Benchmarks for science literacy online. *www.project2061.org/publications/bsl/online*

Driver, R., A. Squires, P. Rushworth, and V. Wood-Robinson. 1994. *Making sense of secondary science: Research into children's ideas.* London: RoutledgeFalmer.

Eisenkraft, A. 2006. ABC: Activity before concept. Presented at the National Science Teachers Association National Conference, St. Louis.

Keeley, P. 2008. *Science formative assessment: 75 practical strategies for linking assessment, instruction, and learning.* Thousand Oaks, CA: Corwin Press and Arlington, VA: NSTA Press

National Council of Teachers of Mathematics (NCTM). 2000. *Principles and standards for school mathematics.* Reston, VA: NCTM.

National Research Council (NRC). 1996. *National science education standards.* Washington, DC: National Academies Press.

Stavy, R., and D. Tirosch. 2000. *How students (mis-) understand science and mathematics: Intuitive rules.* New York: Teachers College Press.

Trowbridge, D., and L. McDermott. 1981. Investigation of student understanding of acceleration in one dimension. *American Journal of Physics* 49 (2): 242–253.

Twigger, D., M. Byard, R. Driver, S. Draper, R. Hartley, S. Hennessy, R. Mohamed, C. O'Malley, T. O'Shea, and E. Scanlon. 1994. The conception of force and motion of students aged between 10 and 15 years: An interview study designed to guide instruction. *International Journal of Science Education* 16(2): 215–229.

Describing Motion and Position

Rolling Marbles

Jen, Debra, and Greg are playing with ramps and marbles. They decide to have a contest to see who can make a marble roll down a ramp the fastest. Each friend uses the same height and identical marbles. They each let go of their marbles at the top of their ramps. (They do not give their marbles a push.)

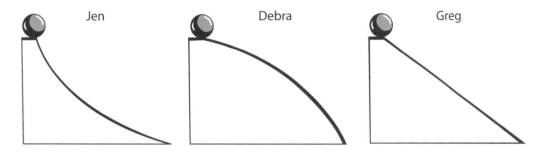

Circle whose marble will reach the bottom of the ramps first.

A Jen's marble

B Debra's marble

C Greg's marble

D No one will win—it will be a tie.

Explain your thinking. Describe your ideas about the time it takes for the marble to reach the end of the different ramps.

_____
_____
_____
_____
_____
_____
_____

Uncovering Student Ideas in Physical Science   59

Rolling Marbles

Teacher Notes

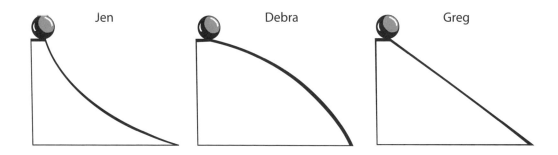

Purpose
The purpose of this assessment probe is to elicit students' ideas about motion using marbles and ramps. The probe is designed to reveal students' thinking about the factors that affect the time it takes for a marble to roll down a ramp.

Related Concepts
acceleration, average speed, speed, time, time interval

Explanation
The best answer is Jen's marble. The marble that rolls down the ramp with the steepest incline at the beginning will get to the bottom first. This is because the average speed will be higher as compared to the other two ramps if the ball is moving faster at the beginning of the motion. A steeper ramp will cause the ball to reach a higher speed in a shorter amount of time as compared to a less steep ramp. A higher average speed means that the ball will get to the bottom of the ramp in a shorter period of time as long as the ramps are of similar length.

Administering the Probe
This probe can be used with elementary, middle school, and high school students. When used with elementary students, the focus should be on observation and comparing the shapes of the ramps, not on the explanation of average speed and acceleration. The setup for this probe can be shown with the props used in the probe—for example, use three pieces of flexible PVC pipe that are the same height and ramp length, identical marbles, and three different ramps or a ramp that can be flexibly bent into the three different shapes. Ask students to predict which marble will reach the bottom of its ramp first.

Describing Motion and Position

Related Ideas in *National Science Education Standards* (NRC 1996)

K–4 Position and Motion of Objects
- An object's motion can be described by tracing and measuring its position over time.

5–8 Motions and Forces
★ The motion of an object can be described by its position, direction of motion, and speed.

Related Ideas in *Benchmarks for Science Literacy* (AAAS 1993, 2009)

K–2 Communication Skills
★ Describe and compare things in terms of number, shape, texture, size, weight, color, and motion.

3–5 Motion
- How fast things move differs greatly.

Related Research
- The words *fast* and *slow* refer to the speed of an object. The words *short* and *long* refer to time intervals. Students often confuse these two related, but different, concepts.
- Naturally, children's ideas and descriptions of motion tend to be less differentiated than those of physicists. Children tend to see objects either at rest or moving; they rarely focus on the period of change. They use everyday terms such as *going faster* in ambiguous ways, sometimes referring to the magnitude of the speed of an object and at other times referring to the speed increasing with time (Driver et al. 1994, p. 155).
- Young children typically start with identifying the direction in which the object moves without regard to the speed of the object. As their ideas progress, they often offer "snapshot" descriptions—that is, a description that is essentially a still photograph of an object, without looking at changes—in which they compare the speed of an object at different locations or instants. Eventually, older children can be led to describe how the speed of an object is changing at a specific location or instant (Dykstra and Sweet 2009).

Suggestions for Instruction and Assessment
- This probe can be used as a P-E-O strategy: commit to a *prediction*, *explain* the reasoning behind the prediction, test the prediction, and *observe* the results. If observations do not match the predictions, students need to rethink their explanations (Keeley 2008). If this probe is used with younger children, explanations should be based on observations of what may have caused one of the marbles to reach the bottom first; they should focus on what is different about the three ramps. Students can further test their explanations by making additional changes to the ramps.
- A second version of this probe can be used as a follow-up with older students. Ask students which marble will be moving the fastest at the end of the ramp. This question is different from the probe, which asks which marble will reach the end of the ramp first. All three marbles will be rolling at approximately the same speed when they reach the bottom of each ramp. This is because the height of all three ramps is the same, so each marble will have the same gravitational potential energy at the top of each ramp. This gravitational potential energy becomes kinetic energy at the bottom of the ramp. If there are no other forces acting on the marbles (i.e., if friction is negligible) then all three marbles will have

★ Indicates a strong match between the ideas elicited by the probe and a national standard's learning goal.

the same kinetic energy at the bottom of each ramp. Because they are identical marbles, they will all have the same speed at the bottom of the ramp.
- Advanced students could further investigate the mathematics behind finding the exact shape of a track that yields the shortest time for a ball to roll. This mathematical curve is called a brachistochrone. A web animation for this curve that provides a representation for the "Rolling Marbles" probe can be found at *http://curvebank. calstatela.edu/brach/brach.htm*.
- Provide younger students with additional opportunities to explore balls and ramps. For example, if the ramps are all the same, does the mass of the ball make a difference? What about the height, length, or material the ramp is made of? Encourage students to make predictions and test them.

References

American Association for the Advancement of Science (AAAS). 1993. *Benchmarks for science literacy.* New York: Oxford University Press.

American Association for the Advancement of Science (AAAS). 2009. *Benchmarks for science literacy online. www.project2061.org/publications/bsl/online*

Driver, R., A. Squires, P. Rushworth, and V. Wood-Robinson. 1994. *Making sense of secondary science: Research into children's ideas.* London: RoutledgeFalmer.

Dykstra, D., and D. Sweet. 2009. Conceptual development about motion and force in elementary and middle school students. *American Journal of Physics* (77) 5: 468–476.

Keeley, P. 2008. *Science formative assessment: 75 practical strategies for linking assessment, instruction, and learning.* Thousand Oaks, CA: Corwin Press and Arlington, VA: NSTA Press.

National Research Council (NRC). 1996. *National science education standards.* Washington, DC: National Academies Press.

Describing Motion and Position

String Around the Earth

Imagine that you tied a string around the center of the Earth along the equator. The string lies on top of the ground and the oceans. You then untied the string and added 6 more meters to the string. You pulled the string away from all sides of the Earth equally. What is the largest animal that could crawl under the string?

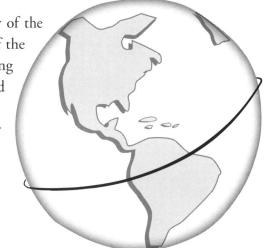

A ant

B mouse

C cat

D goat

E horse

Explain your thinking. Provide an explanation for your answer.

_____
_____
_____
_____
_____
_____
_____
_____
_____

Uncovering Student Ideas in Physical Science

Describing Motion and Position

String Around the Earth

Teacher Notes

Purpose
The purpose of this mathematical thought experiment probe is to elicit students' ideas about a highly counterintuitive ratio problem. Many ideas students encounter in physics do not fit with their commonsense views of the natural world. This adaptation of a timeless, classic mathematics question (Arons 1977; Leiber 1942) helps teachers recognize whether students can reason with proportions.

Related Concepts
circumference, proportion, ratio

Explanation
The best answer is D: goat. The answer is highly counterintuitive. Even after performing the calculation, many students will still not believe the answer. First, one needs to understand that the ratio of circumference to diameter (pi) is true regardless of the size of the circle. Pi is a number that represents the change in the circumference of a circle for each unit of change in the diameter. If the circumference increases by 6 m then the diameter will change by 6/pi m or by about 2 m. Therefore the string will extend about a meter above the Earth on all sides. This answer is independent of the size of the circle that you start with (which is why you do not need to know the circumference of the Earth to answer this question).

This probe is useful in showing how we often rely on our intuitive, commonsense ideas. Commonsense ideas usually have a nugget of truth (e.g., the 6 m addition of string is negligible compared to the circumference of the Earth). However, science and mathematics allow us to refine our commonsense ideas (e.g., the 1 m increase in the radius is indeed negligible compared to the radius of the Earth). What does this question have to do with motion? Many concepts in physics require an understanding of proportions. Pi can be thought of as the increase in the circumference of a circle for every unit of increase in the diameter. Similarly, speed is the distance an object travels for

Describing Motion and Position

each unit of time. Pi is the slope of a graph of circumference versus diameter and speed is the slope of a graph of position versus time.

Administering the Probe

Because of the sophisticated nature of this probe, it is best used with middle or high school students who have the procedural and conceptual mathematical knowledge to perform and understand the calculation. If students are unsure of the height of the animals, you might consider providing height data such as the following: ant is 1 mm or .001 m; mouse is 3 cm or .03 m; cat is 30 cm or .3 m; goat is 90 cm or .9 m; and horse is 2 m. Another alternative is to provide only two choices—the ant and the goat.

Related Ideas in *National Science Education Standards* (NRC 1996)

5–8 Abilities of Inquiry
★ Use mathematics on all aspects of scientific inquiry.

9–12 Abilities of Inquiry
• Use technology and mathematics to improve investigations and communications.

Related Ideas in *Benchmarks for Science Literacy* (AAAS 1993, 2009)

6–8 Numbers
★ Some interesting relationships between two variables include the variables always having the same difference or same ratio.

6–8 Computation and Estimation
★ Calculate the circumference and areas of rectangles, triangles, and circles.

9–12 Computation and Estimation
★ Use ratios and proportions, including constant rates, in appropriate problems.

Related Research

- Early adolescents—as well as many adults—have difficulty with proportional reasoning. The difficulty is influenced by the format of the problem, the numbers used in the problem, the types of ratios used, and the context of the problem (AAAS 1993).
- Different ratio types produce different degrees of difficulty (AAAS 1993). For example, in studies by Heller et al. (1989), speed problems appear to be more difficult than exchange problems. Unfamiliarity with the problem situation causes even more difficulty when it occurs with a difficult ratio type (AAAS 1993, p. 360).
- Misailidou and Williams (2003) found little correlation between a child's computational ability and his or her ability to perform proportional reasoning. This result implies that students should be exposed to situations in which they are required to use proportions. They should not be allowed to use algorithms to perform calculations (which many students will want to do to avoid proportional reasoning).

Suggestions for Instruction and Assessment

- Other adaptations and solutions to this problem can be found by Googling "String Around the Earth Problem."
- If students are still troubled by the answer, you can have them test their ideas by starting with a very small circle of string and adding 6 m to the circumference. They can then measure how much the diameter increases. The fact that this increase in diameter is the same even if the circle started at the Earth's circumference of

---

★ Indicates a strong match between the ideas elicited by the probe and a national standard's learning goal.

about 40,000 kilometers is the foundation for understanding pi as a ratio.
- This probe can also be used after an exercise in which students measure and then graph the circumference and the diameter of a variety of circular objects. Students can then be led to interpret the slope of this graph (pi) as the change in circumference for each unit of diameter (McDermott and the Physics Education Group 1996).

References

American Association for the Advancement of Science (AAAS). 1993. *Benchmarks for science literacy.* New York: Oxford University Press.

American Association for the Advancement of Science (AAAS). 2009. Benchmarks for science literacy online. *www.project2061.org/publications/bsl/online*

Arons, A. 1977. *The various language: An inquiry approach to physical science.* New York: Oxford University Press.

Heller, P., A. Ahlgren, T. Post, M. Behr, and R. Lesh. 1989. Proportional reasoning: The effect of two context variables, rate type, and problem setting. *Journal of Research in Science Teaching* (26): 205–220.

Leiber, L. 1942. *The education of T.C. MITS [the celebrated man in the street]: What modern mathematics means to you.* Brooklyn, NY: Long Island University and the Galois Institute of Mathematics and Art.

McDermott, L. C., and the Physics Education Group at the University of Washington. 1996. *Physics by inquiry.* New York: John Wiley and Sons.

Misailidou, C., and J. Williams. 2003. Diagnostic assessment of children's proportional reasoning. *Journal of Mathematical Behavior* 22: 335–368.

National Research Council (NRC). 1996. *National science education standards.* Washington, DC: National Academies Press.

Section 2
Forces and Newton's Laws

	Concept Matrix 68
	Related Curriculum Topic Study Guides 69
	Related NSTA Press Books, NSTA Journal Articles, and NSTA Learning Center Resources 69
14	Talking About Forces 71
15	Does It Have to Touch? 75
16	Force and Motion Ideas 79
17	Friction 83
18	A World Without Friction 87
19	Rolling to a Stop 91
20	Outer Space Push 95
21	Riding in the Parade 99
22	Spaceships 103
23	Apple in a Plane 107
24	Ball on a String 111
25	Why Things Fall 115
26	Pulling on a Spool 119
27	Lifting Buckets 123
28	Finger Strength Contest 127
29	Equal and Opposite 131
30	Riding in a Car 135

Concept Matrix Probes #14–#30

Forces and Newton's Laws

RELATED CONCEPTS ↓ / PROBES →	#14 Talking About Forces	#15 Does It Have to Touch?	#16 Force and Motion Ideas	#17 Friction	#18 A World Without Friction	#19 Rolling to a Stop	#20 Outer Space Push	#21 Riding in the Parade	#22 Spaceships	#23 Apple in a Plane	#24 Ball on a String	#25 Why Things Fall	#26 Pulling on a Spool	#27 Lifting Buckets	#28 Finger Strength Contest	#29 Equal and Opposite	#30 Riding in a Car
GRADE LEVEL USE →	K–5	2–8	6–12	6–12	6–12	6–12	6–12	6–12	9–12	6–12	3–12	6–12	9–12	9–12	6–12	8–12	6–12
acceleration												X					
action-at-a-distance force		X															
active action			X														
centripetal force										X	X						
changing direction									X								X
circular motion									X		X						
constant speed														X			
contact force		X	X	X						X							
direction of motion			X			X			X								
energy transfer						X											
friction				X	X	X											
gravitational force									X	X	X	X					
interaction			X	X	X			X		X			X	X	X	X	
kinetic friction				X													
mass												X					
net force									X		X	X	X				
Newton's first law					X			X				X	X	X			X
Newton's second law												X	X	X			
Newton's third law							X						X		X	X	
normal force										X							
passive action			X							X						X	
pushes and pulls	X																
rolling friction				X									X				
sliding friction				X													
static friction				X								X					
tension													X				
weight												X					

68 National Science Teachers Association

Forces and Newton's Laws

Related Curriculum Topic Study Guides*

Motion
Forces
Laws of Motion

* Guides will be found in Keeley, P. 2005. *Science Curriculum Topic Study: Bridging the Gap Between Standards and Practice.* Thousand Oaks, CA: Corwin Press and Arlington, VA: NSTA Press. Each Curriculum Topic Study Guide shows the reader how to Identify Adult Content Knowledge, Consider Instructional Implications, Identify Concepts and Specific Ideas, Examine Research on Student Learning, Examine Coherency and Articulation, and Clarify State Standards and District Curriculum.

Related NSTA Press Books, NSTA Journal Articles, and NSTA Learning Center Resources

NSTA Press Books

American Association for the Advancement of Science (AAAS). 2001. *Atlas of science literacy.* Vol. 1. (See "Laws of Motion" map, pp. 62–63.) Washington, DC: AAAS.

Eichinger, J. 2008. Experimenting with force and motion using origami frogs. In *Activities linking science with math: K–4,* 67–78. Arlington, VA: NSTA Press.

Eisenkraft, A., and L. Kirkpatrick. 2006. *Quantoons.* Arlington, VA: NSTA Press.

Horton, M. 2009. *Take-home physics: High impact, low-cost labs.* Arlington, VA: NSTA Press.

Keeley, P. 2005. *Science curriculum topic study: Bridging the gap between standards and practice.* Thousand Oaks, CA: Corwin Press and Arlington, VA: NSTA Press.

Morgan, E., and K. Ansberry. 2005. Sheep in a jeep. In *Picture perfect science lessons: Using children's books to guide inquiry* (grades 3–6), 181–204. Arlington, VA: NSTA Press.

Morgan, E., and K. Ansberry. 2007. Roller coasters. In *More Picture Perfect Science Lessons: Using children's books to guide inquiry, k–4,* 133–146. Arlington, VA: NSTA Press.

Robertson, W. 2002. *Force and motion: Stop faking it! Finally understanding science so you can teach it.* Arlington, VA: NSTA Press.

NSTA Journal Articles

Abisdris, G., and A. Phaneuf. 2007. Using a digital video camera to study motion. *The Science Teacher* (Dec.): 44–47.

Adams, B. 2007. Science shorts: Energy in motion. *Science and Children* (Mar.): 58–60.

Blair, M., and R. M. Peterson. 1999. Physics: The final frontier. *The Science Teacher* 66 (Dec.): 40–43.

Cox, C. 2001. Isaac Newton olympics. *Science Scope* (May): 18–22.

Eisenstein, S. 2008. Increasing the drive of your physics class. *The Science Teacher* (Mar.): 62–66.

Harris, J. 2004. Science 101: Are there different types of force and motion? *Science Scope* (Mar.): 19.

King, K. 2005. Making sense of motion. *Science Scope* (Feb.): 22–26.

Robertson, W., J. Gallagher, and W. Miller. 2004. Newton's first law: Not so simple after all. *Science and Children* 41 (Mar.): 25–29.

Spevak, A. 2008. The art of physics: Using cartooning to illustrate Newton's laws of motion. *The Science Teacher* (Mar.): 44–46.

NSTA Learning Center Resources
NSTA Podcasts:

http://learningcenter.nsta.org/products/podcasts.aspx?lid=hp

Force
Circular Motion
Newton's Second Law
Inertia
Newton's Third Law
Law of Gravity

NSTA SciGuides:

http://learningcenter.nsta.org/products/sciguides.aspx?lid=hp

Force and Motion

NSTA SciPacks:
http://learningcenter.nsta.org/products/scipacks. aspx?lid=hp
Force and Motion

NSTA Science Objects:
http://learningcenter.nsta.org/products/science_ objects.aspx?lid=hp
Force and Motion: Newton's First Law
Force and Motion: Newton's Second Law
Force and Motion: Newton's Third Law

Forces and Newton's Laws

Talking About Forces

Five friends were talking about forces. This is what they said:

Rae: "I think a push is a force and a pull is something else."

Scott: "I think a pull is a force and a push is something else."

Yolanda: "I think a force is either a push or a pull."

Miles: "I think forces are neither pushes nor pulls. I think they are something else."

Violet: "I think pushes and pulls are forces, but there is also another type of force that just holds things in place."

Which friend do you agree with the most? _____

Explain your thinking. Describe what you think a force is.

_____
_____
_____
_____
_____

Talking About Forces

Teacher Notes

Purpose
The purpose of this assessment probe is to elicit beginning ideas about forces. The probe is designed to reveal whether students generally identify forces as pushes and pulls.

Related Concepts
pushes and pulls

Explanation
Yolanda has the best answer: "I think a force is either a push or a pull." By basic definition, a force is a push or pull. Some students might think there is a third type of force, which they sometimes refer to as a "holding force"; it is different from a push or pull and "holds" resting objects in place. For example, they might think that a book resting on a table is being neither pulled on nor pushed. In this instance, the pulling force (gravity) and the pushing force (the upward force exerted by the table on the book) are balanced.

Administering the Probe
This probe is best used at the elementary and middle school levels, although it may be useful in determining whether high school students still believe in a "holding force." Young children are just beginning to learn about forces, and so this probe is particularly helpful because it introduces them to "push and pull" vocabulary as a prerequisite to describing how forces affect motion. With very young children, who may not be ready to understand how an object at rest is affected by forces, you might consider eliminating the last distracter (Violet's response).

Related Ideas in *National Science Education Standards* (NRC 1996)

K–4 Position and Motion of Objects
★ The position and motion of objects can be changed by pushing or pulling. The size of the change is related to the strength of the push or pull.

★ Indicates a strong match between the ideas elicited by the probe and a national standard's learning goal.

Forces and Newton's Laws

Related Ideas in *Benchmarks for Science Literacy* (AAAS 1993, 2009)

K–2 Motion
★ The way to change how something is moving is to give it a push or a pull.

3–5 Motion
- The earth's gravity pulls any object toward it without touching it.
- Without touching them, a magnet pulls on all things made of iron and either pushes or pulls on other magnets.
- Without touching them, material that has been electrically charged pulls on all other materials and may either push or pull other charged materials.

Related Research
- Younger students tend to bring lay meanings of the word *force* to their learning. They often associate the word *force* with coercion, living things, physical activity, and muscular strength (Driver et al. 1994).
- Some students have difficulty associating manifestations of force with pushes or pulls (Shevlin 1989; Erickson and Hobbs 1978). For example, some primary-age children did not associate a kick or throw with a push. Students ages 6–14 thought there was a difference between forces that pull and those that just "hold" (Driver et al. 1994).
- Many students widely regard rest as a natural state in which no forces act on an object (Minstrell 1982). Students who recognize there is some type of "holding force" that keeps an object stationary tend to think of it as quite different from a pushing or pulling force (Driver et al. 1994).

Suggestions for Instruction and Assessment
- Take students on a "push and pull" walk. Identify all the examples of things they push or pull or see being pushed or pulled.
- Present students with examples of forces and ask them to decide which examples are pulls and which are pushes. This can be done as a card sort activity.
- Encourage students to come up with their own lists of pushes and pulls.
- Younger students should first have opportunities to experience pushes and pulls with contact forces (a contact force is a force either between two objects or between an object and a surface that are in contact with each other). Later, when they develop the idea of forces that act at a distance, they can explore pushes and pulls with falling objects, magnets, and electrically charged objects.
- Build a word wall of force- and motion-related words. Include the words *push* and *pull* and encourage students to use these words when they describe forces and motions.
- The forces that young children are most familiar with are the ones exerted by their own muscles. When introducing the notion of force as a push or pull, have children push and pull on objects using their own muscles. Then transition to discussing pushes and pulls exerted by other living things and nonliving objects so that the children don't equate force with only human action.
- Bridging analogies (Clement 1993) are useful with middle school students, who may have difficulty accepting the idea that an object at rest, such as a book on a table, has both a pushing and pulling force acting on

---

★ Indicates a strong match between the ideas elicited by the probe and a national standard's learning goal.

it. A good bridging analogy would be to show the class a book resting on a spring and then a book resting on a springy surface like a piece of foam (Driver et al. 1994).

References

American Association for the Advancement of Science (AAAS). 1993. *Benchmarks for science literacy.* New York: Oxford University Press.

American Association for the Advancement of Science (AAAS). 2009. Benchmarks for science literacy online. *www.project2061.org/publications/bsl/online*

Clement, J. 1993. Using bridging analogies and anchoring intuitions to deal with students' preconceptions in physics. *Journal of Research in Science Teaching* 30 (10): 1241–1257.

Driver, R., A. Squires, P. Rushworth, and V. Wood-Robinson. 1994. *Making sense of secondary science: Research into children's ideas.* London: RoutledgeFalmer.

Erickson, G., and E. Hobbs. 1978. The developmental study of student beliefs about force concepts. Paper presented to the annual convention of the Canadian Society for the Study of Education. Ontario, Canada (June 2).

Minstrell, J. 1982. Explaining the "at rest" condition of an object. *The Physics Teacher* 20: 10–14.

National Research Council (NRC). 1996. *National science education standards.* Washington, DC: National Academies Press.

Shevlin, J. 1989. Children's prior conceptions of force aged 5–11 and their relevance to attainment target 10 of the national curriculum of science. M.Ed. thesis, University of Leeds, UK.

Forces and Newton's Laws

Does It Have to Touch?

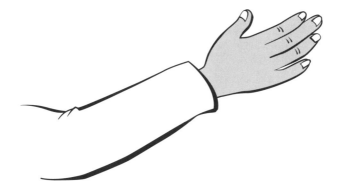

Two friends are arguing about forces. They disagree about whether something has to be touched in order for a force to act. This is what they say:

Akiko: "I think two things have to touch in order to have a force between them."

Fern: "I don't think two things have to touch in order to have a force between them."

Which friend do you most agree with? _____

Explain your thinking. Provide examples that support your ideas about forces.

_____
_____
_____
_____
_____
_____
_____
_____
_____

Uncovering Student Ideas in Physical Science

Forces and Newton's Laws

Does It Have to Touch?

Teacher Notes

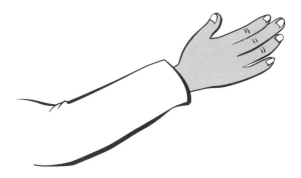

Purpose
The purpose of this assessment probe is to elicit beginning ideas about types of forces. The probe is designed to reveal whether students recognize that forces can act both in direct contact with an object and at a distance.

Related Concepts
action-at-a-distance force, contact force

Explanation
Fern has the best answer: "I don't think two things have to touch in order to have a force between them." Forces can act between objects without the objects being in contact with each other. Forces can involve direct contact between objects (such as a person pulling a wagon or pushing a cart) or action at a distance (such as a ball being pulled toward the Earth by gravity, a magnet attracting or repelling another magnet, or two electrically charged balloons attracting or repelling each other). The action-at-a-distance forces can also act when they are in contact with an object, such as a ball resting on the ground, a nail stuck to a magnet, or a piece of paper stuck to an electrically charged balloon.

Administering the Probe
This probe is best used in the elementary grades after students have developed the concept of force as a push or pull. It can be used when students have learned about forces they have experienced through direct contact and are ready to learn about different types of forces that act at a distance, such as gravity, magnetic force, and electrical force. If needed, clarify what is meant by *touched* in the probe. In this context *touch* means direct, physical contact between two things, such as kicking a ball. The probe is also useful in uncovering middle school students' preconceptions about forces.

Forces and Newton's Laws

Related Ideas in *National Science Education Standards* (NRC 1996)

K–4 Position and Motion of Objects
- The position and motion of objects can be changed by pushing or pulling.

K–4 Light, Heat, Electricity, and Magnetism
- Magnets attract and repel each other and certain kinds of materials.

Related Ideas in *Benchmarks for Science Literacy* (AAAS 1993, 2009)

K–2 Forces of Nature
- Things near the earth fall to the ground unless something holds them up.
- ★ Magnets can be used to make some things move without being touched.

3–5 Forces of Nature
- ★ The earth's gravity pulls any object toward it without touching it.
- ★ Without touching them, a magnet pulls on all things made of iron and either pushes or pulls on other magnets.
- ★ Without touching them, material that has been electrically charged pulls on all other materials and may either push or pull other charged materials.

6–8 Forces of Nature
- The sun's gravitational pull holds the earth and other planets in their orbits, just as the planets' gravitational pull keeps their moons in orbit around them.
- Electric currents and magnets can exert a force on each other.
- A charged object can be charged in one of two ways, which we call either positively charged or negatively charged. Two objects that are charged in the same manner exert a force of repulsion on each other, while oppositely charged objects exert a force of attraction on each other.

Related Research
- Younger students tend to bring lay meanings of the word *force* to their learning. They often associate the word *force* with coercion, living things, physical activity, and muscular strength (Driver et al. 1994).
- Elementary students typically do not understand gravity as a force. They see "falling" as a natural action by an object (AAAS 1993).
- Students often describe electric and magnetic interactions without using words that distinguish among forces, properties of matter, and fields. This is an example of a common student difficulty related to the difference between a cause and an effect (Andersson 1985).
- A study of children ages 3–9 (Selman et al. 1982) found two different conceptual levels of thinking about magnetism. At the first level, children appeared to be simply linking events, without the notion of force being involved. At the second level, children demonstrated an emerging notion of an "unseen force" and they began to talk about a magnet "pulling on things" (Driver et al. 1994).

Suggestions for Instruction and Assessment
- Present students with examples of forces and ask them to decide which ones are pulls and which are pushes. Then ask them to decide which ones involve contact between objects and which do not.
- Younger students should first have opportunities to experience pushes and pulls with contact forces. Later, once they understand the idea of forces that act at

---

★ Indicates a strong match between the ideas elicited by the probe and a national standard's learning goal.

a distance, they can explore pushes and pulls with falling objects, magnets, and electrically charged objects.
- Because gravity, magnetic force, and the force between electrically charged objects cannot be seen, children need to grasp the concept of action at a distance by observing its effects. Give students multiple opportunities to see these effects and to describe them as pushes or pulls.
- Students need to recognize that with the exception of gravitational, magnetic, and electric forces, no other forces act at a distance.

References

American Association for the Advancement of Science (AAAS). 1993. *Benchmarks for science literacy.* New York: Oxford University Press.

American Association for the Advancement of Science (AAAS). 2009. Benchmarks for science literacy online. *www.project2061.org/publications/bsl/online*

Andersson, B. 1985. *The experiential gestalt of causation: A common core to pupils' preconceptions in science.* Gothenburg, Sweden: University of Gothenburg, Department of Education and Educational Research.

Driver, R., A. Squires, P. Rushworth, and V. Wood-Robinson. 1994. *Making sense of secondary science: Research into children's ideas.* London: RoutledgeFalmer.

National Research Council (NRC). 1996. *National science education standards.* Washington, DC: National Academies Press.

Selman, R., M. Krupa, C. Stone, and D. Jacquette. 1982. Concrete operational thoughts and the emergence of the concept of unseen force in children's theories of electromagnetism and gravity. *Science Education* 66 (2): 181–194.

Forces and Newton's Laws

Force and Motion Ideas

Mrs. Li's students share their ideas about force and motion. Here are some of the ideas they come up with. Put an X next to each of the ideas you agree with.

____ **A** If there is motion, then a force is acting.

____ **B** If there is no motion, then there is no force acting.

____ **C** There cannot be a force without motion.

____ **D** Objects can continue moving in a straight line without applying force.

____ **E** When an object is moving, there is always a force in the direction of its motion.

____ **F** Moving objects stop when their force is used up.

____ **G** Forces act on objects at rest.

____ **H** The stronger the force, the faster an object moves.

____ **I** Constant speed results from constant force.

____ **J** A force is necessary in order to change the direction of motion.

____ **K** Forces make things go, losing energy makes them stop.

____ **L** Force can be transferred from one object to another during motion.

Explain your thinking. Summarize your own ideas about force and motion.

_____

_____

_____

_____

Uncovering Student Ideas in Physical Science

Force and Motion Ideas

Teacher Notes

Purpose
The purpose of this assessment probe is to comprehensively elicit students' ideas about the relationship between force and motion. The list of possible answers includes several distracters that are based on learning research; thus the probe will tell you whether your students hold any of these research-identified, commonly held (incorrect) ideas about force and motion.

Related Concepts
active action, constant speed, direction of motion, interaction, passive action

Explanation
The only correct statements are D, G, and J. Statement D comes from Newton's first law of motion. Statement G helps to emphasize the difference between force and an unbalanced force—that is, that objects at rest will often have several forces acting on them, but these forces are balanced. Statement J comes from Newton's second law of motion. If an object is not moving in a straight line then there must be a net force acting on that object. An example is the Moon orbiting around the Sun, which happens because the Sun is exerting a gravitational force on the Moon.

Most of the statements (A, B, C, E, F, H, I, and K) are examples of what is often called "Aristotelian thinking." This type of thinking results from the belief that motion requires an unbalanced force. This idea was first rejected by Galileo, who first proposed constant motion as the natural motion. Later, Isaac Newton realized that it is only a *change* in motion that requires an unbalanced force and Newton was able to quantify this relationship. Many students use the word *force* as something an object possesses—for example, "it has a lot of force" or "may the force be with you." These students may be thinking of force as a property of an object rather than as an interaction between objects. Choosing statements K and L may be indications of this type of thinking.

Forces and Newton's Laws

Administering the Probe
This probe is best used with middle school and high school students. (It can be modified for upper elementary students by removing choices that students are not yet ready to explain.) The probe can be administered as a card sort by writing each of the statements on cards (Keeley 2008). Students then sort the cards into three separate piles: statements they agree with, those they disagree with, and those they are not sure about.

Related Ideas in *National Science Education Standards* (NRC 1996)

K–4 Position and Motion of Objects
- The position and motion of objects can be changed by pushing or pulling. The size of the change is related to the strength of the push or pull.

5–8 Motions and Forces
- ★ An object that is not being subjected to a force will continue to move at a constant speed and in a straight line.
- ★ If more than one force acts on an object along a straight line, then the forces will reinforce or cancel one another. Unbalanced forces will cause changes in the speed or the direction of an object's motion.

9–12 Motions and Forces
- ★ Objects change their motion only when a net force is applied.

Related Ideas in *Benchmarks for Science Literacy* (AAAS 1993, 2009)

K–2 Motion
- The way to change how something is moving is to give it a push or a pull.

3–5 Motion
- Changes in speed or direction of motion are caused by forces.
- The greater the force is, the greater the change in motion will be. The more massive an object is, the less effect a given force will have.

6–8 Motion
- ★ An unbalanced force acting on an object changes its speed or direction of motion, or both.

9–12 Motion
- ★ Any object maintains a constant speed and direction of motion unless an unbalanced outside force acts on it.

Related Research
- Many students think that if an object is moving, then there is a force acting on it. There is a strong belief that a force must be constantly applied in order for motion, including constant speed, to continue (Gunstone and Watts 1985).
- Some students tend to think of force as a property of an object, rather than an interaction between objects (Brown and Clement 1989; Dykstra, Boyle, and Monarch 1992).
- Some students think that forces get things moving but do not stop things (Minstrell 1989). Some students think things stop when the force or energy in the object runs out (Driver et al. 1994).
- A common belief among students of all ages is that all objects eventually slow down and stop (Driver et al. 1994).
- Some students think force is transferred from one object to another (Brown and Clement 1989).

★ Indicates a strong match between the ideas elicited by the probe and a national standard's learning goal.

Suggestions for Instruction and Assessment

- Teachers and researchers have developed several strategies to help students develop an understanding of forces. One strategy, developed by Camp and Clement (1994), is to use "bridging analogies." This strategy involves starting with an "anchoring example" and then extending student ideas toward a "target problem." For example, to introduce gravity ideas, the teacher has students examine the force exerted by various numbers of rubber bands stretched between their fingers (the anchor) to model the mass dependency of the gravitational force (the target). The interaction between each finger is modeled as a single rubber band. The number of fingers represents the amount of mass and the number of rubber bands represents the total force. The total number of rubber bands depends on the number of fingers on each hand (just as the total gravitational force depends on the amount of both masses). For example, three fingers of one hand interacting with two fingers on the other hand would require a total of six rubber bands. The gravitational force exerted by a mass of 2 kg with a mass of 3 kg is proportional to the masses multiplied together.
- Most approaches to teaching about force require students to have a firm understanding of kinematics (i.e., describing the motion of objects without considering the causes leading to the motion) and be able to identify changes in motion (acceleration). However, one interesting alternative is to introduce energy ideas *before* the study of motion. Students then observe different types of motion and infer changes in energy. An example of the use of this approach is found in *Interactions in Physical Science* (Goldberg et al. 2009).

References

American Association for the Advancement of Science (AAAS). 1993. *Benchmarks for science literacy.* New York: Oxford University Press.

American Association for the Advancement of Science (AAAS). 2009. Benchmarks for science literacy online. *www.project2061.org/publications/bsl/online*

Brown, D., and J. Clement. 1989. Overcoming misconceptions via analogical reasoning: Abstract transfer versus explanatory model construction. *Instructional Science* 18: 237–261.

Camp, C., and J. Clement. 1994. *Preconceptions in mechanics: Lessons dealing with students' conceptual difficulties.* Dubuque, IA: Kendall Hunt.

Driver, R., A. Squires, P. Rushworth, and V. Wood-Robinson. 1994. *Making sense of secondary science: Research into children's ideas.* London: RoutledgeFalmer.

Dykstra, D., C. Boyle, and I. Monarch. 1992. Studying conceptual change in learning physics. *Science Education* 76 (6): 615–652.

Goldberg, F., B. Bendall, P. Heller, and R. Poel. 2009. *Interactions in physical science.* Armonk, NY: It's About Time.

Gunstone, R., and M. Watts. 1985. Force and motion. In *Children's ideas in science*, ed. R. Driver, E. Guesne, and A. Tiberghien, 85–104. Milton Keynes, UK: Open University Press.

Keeley, P. 2008. *Science formative assessment: 75 practical strategies for linking assessment, instruction, and learning.* Thousand Oaks, CA: Corwin Press and Arlington, VA: NSTA Press.

Minstrell, J. 1989. Teaching science for understanding. In *Toward the thinking curriculum: Current cognitive research*, ed. L. Resnick and L. Klopfer, 129–149. Alexandria, VA: Association for Supervision and Curriculum Development.

National Research Council (NRC). 1996. *National science education standards.* Washington, DC: National Academies Press.

Forces and Newton's Laws

Friction

Ansel slides down the playground slide. He notices that the new pants he is wearing slow him down on the slide. When he mentions this to his friend Sergio, Sergio says the rubbing of Ansel's pants against the slide was caused by something called friction—and it is the friction that made Ansel slow down. Ansel wonders about the other ways friction could occur. Check off all the kinds of contact that could cause friction.

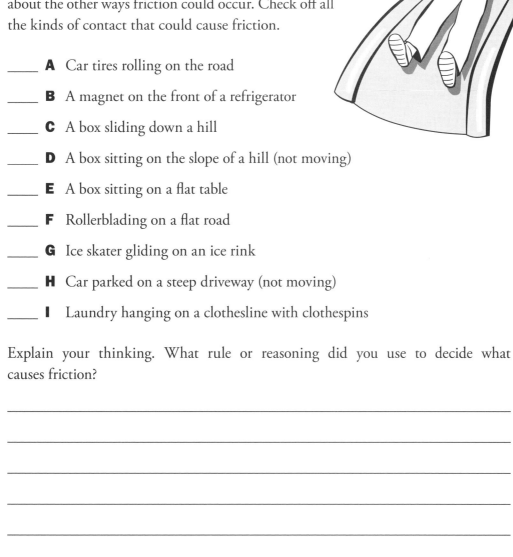

____ **A** Car tires rolling on the road

____ **B** A magnet on the front of a refrigerator

____ **C** A box sliding down a hill

____ **D** A box sitting on the slope of a hill (not moving)

____ **E** A box sitting on a flat table

____ **F** Rollerblading on a flat road

____ **G** Ice skater gliding on an ice rink

____ **H** Car parked on a steep driveway (not moving)

____ **I** Laundry hanging on a clothesline with clothespins

Explain your thinking. What rule or reasoning did you use to decide what causes friction?

_____
_____
_____
_____
_____
_____

Uncovering Student Ideas in Physical Science

Forces and Newton's Laws

Friction

Teacher Notes

Purpose
The purpose of this assessment probe is to elicit students' ideas about friction between solid objects. The probe is designed to determine (a) whether students recognize friction as an interaction between two objects or materials that rub against/slide over each other and (b) whether they limit this interaction to a particular type of matter or contact.

Related Concepts
contact force, friction, interaction, kinetic friction, rolling friction, sliding friction, static friction

Explanation
All of the choices are examples of friction with the exception of E (box sitting on a flat table). A wheel rolling on a surface is an example of "rolling friction." A magnet is attracted to the refrigerator with a magnetic force, but it is the frictional force that prevents the magnet from sliding down the front of the refrigerator. A box would slide down a hill if there were no friction between the box and the hill. Although an ice rink provides a low friction surface, there is still some friction between an ice skate and the ice rink (the skater will eventually slow down and stop if she does not push off). Even a clothespin can only work because of the friction force between the clothes and the surface of the clothespin. In all of the examples except for A and F, which are examples of rolling friction, the friction force acts parallel to the two surfaces that are in contact with each other. Because the surfaces are sliding past each other, C and G are examples of what is called "kinetic friction." The other examples (except for A and F) are called "static friction" because the surfaces are not moving relative to one another.

Administering the Probe
This probe is best used with middle school and high school students. It can also be used with upper elementary students by removing the

Forces and Newton's Laws

static friction distracters that don't involve a visible rubbing or sliding (i.e., B, D, H, and I).

Related Ideas in *National Science Education Standards* (NRC 1996)

5–8 Motions and Forces
★ Unbalanced forces will cause changes in the speed or direction of an object's motion.

9–12 Motions and Forces
- Objects change their motion only when a net force is applied.

Related Ideas in *Benchmarks for Science Literacy* (AAAS 1993, 2009)

3–5 Motion
- Changes in speed or direction of motion are caused by forces.

6–8 Motion
★ An unbalanced force acting on an object changes its speed or direction of motion, or both.

9–12 Forces of Nature
★ Electric forces hold solid and liquid materials together and act between objects when they are in contact—as in sticking or sliding friction.

Related Research
- In a study by Stead and Osborne (1981), 50% of 13-year-olds in their sample group of 38 12- to 16-year-olds associated friction with rubbing (Driver et al. 1994).
- Some students thought that if a box is motionless on a slope, there is no friction (because there is no rubbing, heat, or wearing down of surfaces) (Stead and Osborne 1980).
- In a group of 47 secondary students, the following ideas about friction were held: Friction occurs only between solids (12 students) and friction occurs with liquids but not with gases (10 students). Only 16 students called friction a force (the responses of 9 students were in an "other" category) (Stead and Osborne 1980).

Suggestions for Instruction and Assessment
- Ask students to recall times when they easily slid down a slide on a playground and other times when they were slowed down on the slide. Ask them to compare the interaction between themselves and the slide that occurred in each case.
- It is important to develop the concept of interactions when teaching about friction.
- Encourage students to come up with examples of friction involving moving and stationary objects.
- Consider extending the probe to examples that include fluid friction, such as the drag on an object in air or water.
- Research shows that some students may benefit from the introduction of an intermediate model of friction, similar to the bridging analogies used by Clement (1993) (Besson and Viennot 2004). For example, have students try to slide two hairbrushes across each other. Students will see the bristles pushing against each other.

References
American Association for the Advancement of Science (AAAS). 1993. *Benchmarks for science literacy.* New York: Oxford University Press.

American Association for the Advancement of Science (AAAS). 2009. Benchmarks for science literacy online. www.project2061.org/publications/bsl/online

Besson, U., and L. Viennot. 2004. Using models at the mesoscopic scale in teaching physics: Two

---

★ Indicates a strong match between the ideas elicited by the probe and a national standard's learning goal.

experimental interventions in solid friction and fluid statics. *International Journal of Science Education* 26 (9): 1083–1110.

Clement, J. 1993. Using bridging analogies and anchoring intuitions to deal with students' preconceptions in physics. *Journal of Research in Science Teaching* 30 (10): 1241–1257.

Driver, R., A. Squires, P. Rushworth, and V. Wood-Robinson. 1994. *Making sense of secondary science: Research into children's ideas.* London: RoutledgeFalmer.

National Research Council (NRC). 1996. *National science education standards.* Washington, DC: National Academies Press.

Stead, K., and R. Osborne. 1980. Friction. LISP working paper 19. Hamilton, New Zealand: University of Waikato, Science Education Research Unit.

Stead, K., and R. Osborne. 1981. What is friction? Some children's ideas. *New Zealand Science Teacher* 27: 51–57.

Forces and Newton's Laws

A World Without Friction

Karen and Jeff walk to school each day. Part of their trip is flat, part is uphill, and part is downhill. Karen and Jeff are trying to imagine what it would be like to walk to school if there were no friction in the world.

Karen: "To go to school in the morning, I could just push off from my front door and I could slide all the way to school!"

Jeff: "No, I don't think that would work. You would eventually slow down after awhile."

Whom do you agree with and why?

_____
_____
_____
_____
_____
_____
_____
_____

Uncovering Student Ideas in Physical Science   87

A World Without Friction

Teacher Notes

Purpose
The purpose of this probe is to examine students' ideas about an imaginary frictionless environment. The probe is designed to reveal students' ideas about the effect of friction on motion.

Related Concepts
friction, Newton's first law

Explanation
The best answer is Karen's: "To go to school in the morning, I could just push off from my front door and I could slide all the way to school!" This is because without friction there is no way to dissipate kinetic energy; therefore you would continue to move in a straight line without slowing down or speeding up unless acted on by another force. (Karen better aim well because once she kicks off, there is no changing direction!) For example, you would slow down if you started to go up a hill (but then you would eventually turn around and speed up in the opposite direction) or you would change direction if you ran into an object.

Administering the Probe
The probe is best used with middle school and high school students to encourage discussion about what a frictionless environment would be like.

Related Ideas in *National Science Education Standards* (NRC 1996)

5–8 Motions and Forces
- Unbalanced forces will cause changes in the speed or direction of an object's motion.

9–12 Motions and Forces
- ★ Objects change their motion only when a net force is applied.

★ Indicates a strong match between the ideas elicited by the probe and a national standard's learning goal.

Forces and Newton's Laws

Related Ideas in *Benchmarks for Science Literacy* (AAAS 1993, 2009)

6–8 Motion
- An unbalanced force acting on an object changes its speed or direction of motion, or both.

9–12 Motion
- Any object maintains a constant speed and direction of motion unless an unbalanced force acts on it.

Related Research
- In a study by Stead and Osborne (1981), friction was associated with rubbing by 50% of 13-year-olds in the sample group of 38 12- to 16-year-olds (Driver et al. 1994).
- Researchers have found that there is a widely held view among students of all ages that there is something called "force" within an object that keeps it moving. Students who have this notion of force (impetus idea) believe a pushed object will eventually slow down and stop because it runs out of "force" or that a force needs to keep being applied in order to maintain constant motion (Driver et al. 1994).
- Most 15-year-olds expect a moving object to come to a stop, even when there is no friction. This explains the student resistance that teachers can face when trying to teach Newton's first law (Driver et al. 1994).

Suggestions for Instruction and Assessment
- Drama and imagination can play an important role in connecting a student's experiences to scientific principles (Wilhelm and Edmiston 1998). What a student can imagine is often an important window into how he or she is thinking. Younger students reveal what they are thinking when we ask them to "act out" a situation. Older students can be asked to write a short story with the title "How I Got to School Today in a World Without Friction" and then share their stories with one another.
- Ask students to come up with examples of friction in their everyday lives and discuss what would happen if those examples became "frictionless."
- Compare and contrast the motion of an object on different surfaces (low friction–high friction; flat; uphill-downhill) when it has been given a push. Then consider what would happen if it were possible to have a frictionless surface and an object were given a push on those three different surfaces.

References

American Association for the Advancement of Science (AAAS). 1993. *Benchmarks for science literacy.* New York: Oxford University Press.

American Association for the Advancement of Science (AAAS). 2009. Benchmarks for science literacy online. *www.project2061.org/publications/bsl/online*

Driver, R., A. Squires, P. Rushworth, and V. Wood-Robinson. 1994. *Making sense of secondary science: Research into children's ideas.* London: RoutledgeFalmer.

National Research Council (NRC). 1996. *National science education standards.* Washington, DC: National Academies Press.

Stead, K., and R. Osborne. 1981. What is friction? Some children's ideas. *New Zealand Science Teacher* 27: 51–57.

Wilhelm, J. D., and B. Edmiston. 1998. *Imagining to learn: Inquiry, ethics, and integration through drama.* Portsmouth, NH: Heinemann.

Forces and Newton's Laws

Rolling to a Stop

Rachel is wearing roller skates and is standing next to a wall. She pushes off from the wall and then glides to a stop. Place an X next to all the statements that you think are true about Rachel's motion.

____ **A** The force from the wall is becoming less and less as Rachel glides on her skates.

____ **B** Rachel's motion energy is naturally going down and the energy is disappearing.

____ **C** Rachel's motion energy is being turned into other types of energy.

____ **D** There is a force by the ground slowing Rachel down.

____ **E** There is a force by the ground that keeps Rachel moving for a while before she stops.

Explain your thinking about why Rachel slowed down and stopped after pushing off from the wall.

_____
_____
_____
_____
_____
_____
_____
_____
_____

Uncovering Student Ideas in Physical Science

Forces and Newton's Laws

Rolling to a Stop

Teacher Notes

Purpose
The purpose of this assessment probe is to elicit student ideas related to force, energy, and motion. The probe is designed to identify students who may think about force as being carried by an object, rather than as being an interaction between objects. The probe will also identify those students who may confuse energy ideas with force ideas.

Related Concepts
energy transfer, friction, interaction

Explanation
The two best answers are (C) Rachel's motion energy is being turned into other types of energy, and (D) There is a force by the ground slowing Rachel down. Changes in energy are evidence of an interaction. An example of an interaction is a contact force between two objects. In this case, there is a force acting by the floor on the skates that is causing Rachel to slow down. This type of force (between a wheel and the surface on which it rolls) is called rolling friction (although one can also consider the rotational friction within the wheels of the skates). There is also an energy transfer from Rachel and her skates to the floor, as well as motion energy that is transferred to the wheels (in the form of heat) due to the friction within the wheels. The floor and the rubbing in the wheels are taking energy away as Rachel moves on her skates. Distracter A implies that the force exerted by the wall is transferred and carried by Rachel. Distracter B is partially correct in that Rachel's motion energy is decreasing; however, the motion energy does not disappear—it is transferred to other locations, including the floor, the air, and the skates themselves.

Administering the Probe
This probe is best used with middle school and high school students. If roller skates or a skateboard is available, this probe can be modeled in the classroom. (If teachers prefer to use a *sliding friction* example, such as ice skates,

Forces and Newton's Laws

instead of *rolling friction*—which is used in this example because students can use roller skates or a skateboard on the classroom floor whereas an ice rink would be difficult to duplicate in the classroom—the probe can be modified accordingly.)

Related Ideas in *National Science Education Standards* (NRC 1996)

5–8 Motions and Forces
- An object that is not being subjected to a force will continue to move at a constant speed and in a straight line.
- ★ If more than one force acts on an object along a straight line, then the forces will reinforce or cancel one another. Unbalanced forces will cause changes in the speed or the direction of an object's motion.

5–8 Transfer of Energy
- ★ Energy is transferred in many ways.

9–12 Motions and Forces
- Objects change their motion only when a net force is applied.

Related Ideas in *Benchmarks for Science Literacy* (AAAS 1993, 2009)

3–5 Motion
- Changes in speed or direction of motion are caused by forces.

6–8 Motion
- ★ An unbalanced force acting on an object changes its speed or direction of motion, or both.

6–8 Energy Transformations
- Whenever energy appears in one place, it must have disappeared from another. Whenever energy is lost from somewhere, it must have gone somewhere else.
- ★ Energy appears in different forms and can be transformed within a system. Motion energy is associated with the speed of an object.

9–12 Motion
- Any object maintains a constant speed and direction of motion unless an unbalanced outside force acts on it.

Related Research
- Many students think that if an object is moving, then there is a force acting on it. There is a strong belief that a force must be constantly applied in order for motion, including constant speed, to continue (Gunstone and Watts 1985).
- Some students tend to think of force as a property of an object, rather than an interaction between objects (Brown and Clement 1989; Dykstra, Boyle, and Monarch 1992).
- Some students think forces only get things moving but are not responsible for stopping things (Minstrell 1989). Some think that things stop when the force or energy in the object runs out (Driver et al. 1994).
- A common belief among students of all ages is that all objects eventually slow down and stop (Driver et al. 1994).

Suggestions for Instruction and Assessment
- This probe can be used prior to instruction to elicit student ideas about the words *force* and *energy*. Many students come to class with a sophisticated vocabulary but without an understanding of what these words mean. A full development of these ideas should include the introduction of the principle of the conservation of energy.

★ Indicates a strong match between the ideas elicited by the probe and a national standard's learning goal.

Uncovering Student Ideas in Physical Science

- When used after instruction, the results from this probe can help teachers better understand student difficulties that persist. Teachers can then revise their instruction accordingly. They can ask the students to describe all the forces that are acting on the skater (a) during the push against the wall and (b) after the skater has let go. Also, teachers can ask students to consider all energy transfers and transformations during parts (a) and (b).

References

American Association for the Advancement of Science (AAAS). 1993. *Benchmarks for science literacy.* New York: Oxford University Press.

American Association for the Advancement of Science (AAAS). 2009. Benchmarks for science literacy online. *www.project2061.org/publications/bsl/online*

Brown, D., and J. Clement. 1989. Overcoming misconceptions via analogical reasoning: Abstract transfer versus explanatory model construction. *Instructional Science* 18: 237–261.

Driver, R., A. Squires, P. Rushworth, and V. Wood-Robinson. 1994. *Making sense of secondary science: Research into children's ideas.* London: RoutledgeFalmer.

Dykstra, D., C. Boyle, and I. Monarch. 1992. Studying conceptual change in learning physics. *Science Education* 76 (6): 615–652.

Gunstone, R., and D. Watts. 1985. Force and motion. In *Children's ideas in science,* ed. R. Driver, E. Guesne, and A. Tiberghien, 85–104. Milton Keynes, UK: Open University Press.

Minstrell, J. 1989. Teaching science for understanding. In *Toward the thinking curriculum: Current cognitive research*, ed. L. Resnick and L. Klopfer, 129–149. Alexandria, VA: Association for Supervision and Curriculum Development.

National Research Council (NRC). 1996. *National science education standards.* Washington, DC: National Academies Press.

Forces and Newton's Laws

Outer Space Push

A box is lying on the table. You give the box a quick shove and notice that the box slides on the table for a short time and then comes to a stop. You then do the same thing on a smooth floor. With the same push from your hand, the box slides for a longer time, but then eventually comes to a stop. You wonder what would happen if you could push the box in outer space, away from any other planets or atmosphere. If

you could give the box the same push, what do you think would happen? Circle the answer that best matches your thinking.

A The box will move forever because nothing is slowing the box down.

B The box will slow down because the push that you gave it will eventually wear out.

C The box will slow down because it will eventually lose all its energy.

Explain your thinking. Describe the reasoning you used for your answer.

_____
_____
_____
_____
_____
_____
_____

Uncovering Student Ideas in Physical Science

Forces and Newton's Laws

Outer Space Push
Teacher Notes

Purpose
This probe is a type of thought experiment designed to elicit students' ideas about Newton's first law of motion. It is designed to reveal whether students recognize that an object in motion will continue moving unless acted upon by a force.

Related Concepts
Newton's first law

Explanation
The best answer is A: The box will move forever because nothing is slowing the box down. If there are no forces acting on the box, then the box will continue to move in a straight line at constant speed. This is consistent with Newton's first law of motion, which states that objects in motion tend to stay in motion unless acted upon by a force. Because there is no outside force acting on the moving box, it will stay in motion in outer space.

Administering the Probe
This probe is best used in middle school and high school. Make sure students understand the context of outer space before asking them to respond to the probe.

Related Ideas in *National Science Education Standards* (NRC 1996)

5–8 Motions and Forces
★ An object that is not being subjected to a force will continue to move at a constant speed and in a straight line.

9–12 Motions and Forces
• Objects change their motion only when a net force is applied.

★ Indicates a strong match between the ideas elicited by the probe and a national standard's learning goal.

Forces and Newton's Laws

Related Ideas in *Benchmarks for Science Literacy* (AAAS 1993, 2009)

3–5 Motion
- Changes in speed or direction of motion are caused by forces.

6–8 Motion
- An unbalanced force acting on an object changes its speed or direction of motion, or both.

9–12 Motion
★ Any object maintains a constant speed and direction of motion unless an unbalanced outside force acts on it.

Related Research
- Researchers have found that there is a widely held view among students of all ages that there is something called "force" within an object that keeps it moving. Students who have this notion of force (impetus idea) believe that a pushed object will eventually slow down and stop because it runs out of "force" or that a force needs to keep being applied in order to maintain constant motion (Driver et al. 1994).
- Most 15-year-olds expect a moving object to come to a stop, even when there is no friction. This explains the student resistance that teachers can face when trying to teach Newton's first law (Driver et al. 1994).
- Some students believe that inertia is an intrinsic resistance of an object to motion (Halloun and Hestenes 1985).

Suggestions for Instruction and Assessment
- It is difficult for students to imagine a world where no forces are acting, but even when they can conceive of such a world, some students will continue to believe that the box will still slow down (because it runs out of either "force" or "energy"). These students are using reasoning that is consistent with what is often called impetus theory, which was first put forward by Aristotle. Listen to how students use the word *force* in a sentence. Students who say that "the book now has a force" are probably thinking of force as something that the hand gives the box and that the box then carries as it moves. Instead, they need to understand that force is an interaction between the hand and the box that only occurs while the hand is in contact with the box.
- The idea that an object in motion will stay in motion unless acted on by a force is difficult to demonstrate because of the presence of friction. Science supply companies make "low friction" carts, air tables, air tracks, and dry ice pucks to help students understand this idea. An inexpensive "hovercraft" can be made from a balloon, a bottle cap, and an old CD. A Google search of "CD Hovercraft" will turn up several websites that explain how to make one of these devices.

References
American Association for the Advancement of Science (AAAS). 1993. *Benchmarks for science literacy.* New York: Oxford University Press.

American Association for the Advancement of Science (AAAS). 2009. Benchmarks for science literacy online. *www.project2061.org/publications/bsl/online*

Driver, R., A. Squires, P. Rushworth, and V. Wood-Robinson. 1994. *Making sense of secondary science: Research into children's ideas.* London: RoutledgeFalmer.

Halloun, I., and D. Hestenes. 1985. Common sense concepts about motion. *American Journal of Physics* 53 (11): 1056–1065.

National Research Council (NRC). 1996. *National science education standards.* Washington, DC: National Academies Press.

★ Indicates a strong match between the ideas elicited by the probe and a national standard's learning goal.

Forces and Newton's Laws

Riding in the Parade

Cindy is very excited—she has been asked to ride in a parade! During the parade, she stands in the middle of the float and waves to the crowd while the float is moving down the street at a constant speed. While she is waving, she sees her friend standing on the sidewalk. She jumps straight up as high as she can so that her friend will see her.

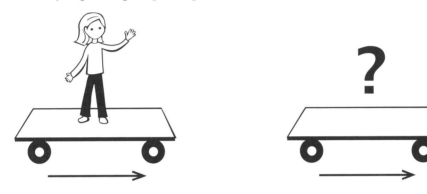

When Cindy lands back on the float, where will she land?

A She will land in the same place on the float from where she jumped.

B She will land closer to the front of the float than where she was before she jumped.

C She will land closer to the back of the float than where she was before she jumped.

Draw where Cindy will land on the float (below).

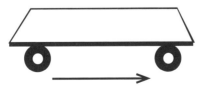

Explain your thinking. Describe the reasoning you used to make your prediction.

_____
_____

Uncovering Student Ideas in Physical Science

Riding in the Parade

Teacher Notes

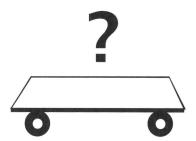

Purpose
The purpose of this probe is to elicit students' ideas about relative motion. The probe is designed to reveal whether students use Newton's first law of motion to predict where a person would land on a moving object if he or she jumped straight up while the object was moving.

Related Concepts
Newton's first law

Explanation
The best answer is A: She will land in the same place on the float from where she jumped—that is, she will land in the middle of the float. She and the float are both moving with the same speed. She will continue her forward motion even though she is no longer in contact with the float. Another example of this motion is when a person jumps up inside an airplane that is moving very fast. This person will land back down directly where he or she jumped from, even though the airplane is moving while the person is up in the air. Newton's first law says that objects in motion continue in motion unless acted on by a force. The float pushes Cindy up to allow her to jump. Nothing is pushing Cindy sideways, so her sideways motion will continue to be the same, that is, she will have a steady sideways speed. More advanced students may try to take air resistance into account and will predict that she lands slightly in back of the center line. This is also a correct statement if accompanied by the correct explanation.

Administering the Probe
This probe is best used with middle school and high school students. Be sure students know that the float is not speeding up, slowing down, or changing direction. It is moving at a steady speed in one direction.

Forces and Newton's Laws

Related Ideas in *National Science Education Standards* (NRC 1996)

5–8 Motions and Forces
★ An object that is not being subjected to a force will continue to move at a constant speed and in a straight line.

9–12 Motions and Forces
★ Objects change their motion only when a net force is applied.

Related Ideas in *Benchmarks for Science Literacy* (AAAS 1993, 2009)

3–5 Motion
- Changes in speed or direction of motion are caused by forces.

6–8 Motion
★ An unbalanced force acting on an object changes its speed or direction of motion, or both.

9–12 Motion
★ Any object maintains a constant speed and direction of motion unless an unbalanced outside force acts on it.

Related Research
- Everyday experiences from birth onward result in firmly established ideas called "gut dynamics." These "gut dynamics" underlie most people's ability to interact with moving objects and to play sports. In addition, people appear to generate for themselves a set of explanations and rules for why things move the way they do. These rules have been termed "lay dynamics" (Driver et al. 1994, p. 154).
- Several researchers have found that computer simulations can help students to understand relative motion (Monaghan and Clement 2000; Morecraft 1985).

Suggestions for Instruction and Assessment
- Students can sit on a rolling skateboard or other moving object and throw a ball straight up into the air to see what happens. (*Note:* It is difficult to toss a ball straight up. Before sitting on a rolling skateboard or other moving object in the classroom, have students practice "pushing" the ball off their palms so it rises up slightly in a straight line and lands back on their palms.) Some science supply companies sell a cart that contains a spring-loaded launcher that will shoot a ball straight up into the air. This common demonstration shows that when the cart is moving at constant speed in a straight line, the ball will land directly back into the launcher.
- A useful resource is the DVD movie collection called *Physics Cinema Classics*. These film clips show a wide variety of motions, including visual evidence of Newton's first law. The DVD can be ordered from the American Association of Physics Teachers: *www.aapt.org/Store/cinemaclassics.cfm*.

References
American Association for the Advancement of Science (AAAS). 1993. *Benchmarks for science literacy.* New York: Oxford University Press.

American Association for the Advancement of Science (AAAS). 2009. Benchmarks for science literacy online. *www.project2061.org/publications/bsl/online*

Driver, R., A. Squires, P. Rushworth, and V. Wood-Robinson. 1994. *Making sense of secondary science: Research into children's ideas.* London: RoutledgeFalmer.

★ Indicates a strong match between the ideas elicited by the probe and a national standard's learning goal.

Monaghan, J. M., and J. Clement. 2000. Algorithms, visualization, and mental models: High school students' interactions with a relative motion simulation. *Journal of Science Education and Technology* 9: 311–325.

Morecroft, L. E. 1985. A relative-motion microworld. Technical Report 347. Cambridge, MA: Massachusetts Institute of Technology.

National Research Council (NRC). 1996. *National science education standards.* Washington, DC: National Academies Press.

Forces and Newton's Laws

Spaceships

You are part of a team that has been asked to choose a design for a plane for outer space. It is very important that the plane be able to turn quickly. The engines that are attached to each plane are shown with this symbol: ⬠

Circle the design that will BEST allow the plane to turn quickly in space.

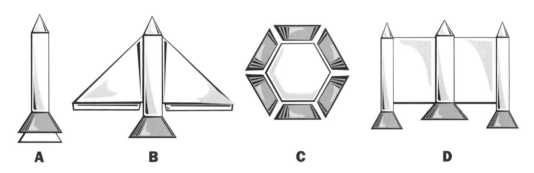
A   **B**   **C**   **D**

Explain your thinking. Describe the reasoning that helped you decide which design to select.

_____
_____
_____
_____
_____
_____
_____
_____
_____
_____
_____

Uncovering Student Ideas in Physical Science

Forces and Newton's Laws

Spaceships
Teacher Notes

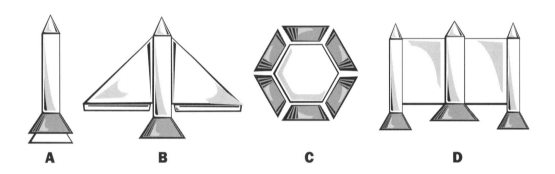

A　　　　B　　　　C　　　　D

Purpose
This probe is designed to elicit students' ideas about changing the direction of motion in the absence of air. Many students will have seen movies or television shows in which spaceships turn by banking or using wing flaps. In outer space, where there is no air to push off of, these motions would not be possible.

Related Concepts
changing direction, direction of motion, Newton's first law

Explanation
Design C—the hexagon shape (with multiple engines)—is the only design that would be able to turn in space. There is no atmosphere in space, so a plane cannot turn by banking or using "flaps" on the wings. With engines that point in only one direction, a plane would be able to speed up only in that direction and would not be able to slow down or turn. Students who have seen science fiction movies such as *Star Wars* may have been misled by the incorrect design of spacecraft used in those types of movies. The hexagon shape has engines that face in many directions, allowing the plane to speed up (in any direction toward which an engine is pointing) and to slow down by firing an engine that will push in the opposite direction of motion.

Administering the Probe
This probe is best used with upper middle school and high school students. Before using it, make sure that students understand that in space, where the plane will be flying, there is no air or atmosphere.

Forces and Newton's Laws

Related Ideas in *National Science Education Standards* (NRC 1996)

5–8 Motions and Forces
- An object that is not being subjected to a force will continue to move at a constant speed and in a straight line.

9–12 Motions and Forces
★ Objects change their motion only when a net force is applied.

Related Ideas in *Benchmarks for Science Literacy* (AAAS 1993, 2009)

3–5 Motion
- Changes in speed or direction of motion are caused by forces.

6–8 Motion
★ An unbalanced force acting on an object changes its speed or direction of motion, or both.

9–12 Motion
★ Any object maintains a constant speed and direction of motion unless an unbalanced outside force acts on it.

Related Research
- Students need numerous opportunities to apply Newton's three laws to experiences outside the classroom (Pugh 2004).
- Many students have a deep-seated belief that motion implies force—an idea that is based on their everyday experiences in an environment with friction. This belief can hinder their understanding of Newton's first and second laws, which describe the relationship between force and acceleration (or a *change* in motion) (Clement 1982).

Suggestions for Instruction and Assessment
- The NSTA website has a feature called "Blick on Flicks," in which a physics professor reviews movies for scientific accuracy (see, for example, "Amelia and the Physics of Flight"). This free site is *www.nsta.org/publications/blickonflicks.aspx*.
- With older students, you can also use this probe as an example of Newton's second law of motion, which states that the acceleration of an object is in the direction of the net force acting on the object. Each engine can push on the spaceship in only one direction (forward). This is why you would need multiple engines to turn right or left (or to slow down).
- Following the probe, use the formative assessment classroom technique (FACT) called "Misrepresentation Analysis" (Keeley 2008) to have students critique movies such as *Star Wars* that depict spaceships shaped like distractor examples B or D banking sharply to the left or right.

References

American Association for the Advancement of Science (AAAS). 1993. *Benchmarks for science literacy.* New York: Oxford University Press.

American Association for the Advancement of Science (AAAS). 2009. Benchmarks for science literacy online. *www.project2061.org/publications/bsl/online*

Clement, J. 1982. Students' preconceptions in introductory mechanics. *American Journal of Physics* 50: 66–71.

Driver, R., A. Squires, P. Rushworth, and V. Wood-Robinson. 1994. *Making sense of secondary science: Research into children's ideas.* London: RoutledgeFalmer.

Keeley, P. 2008. *Science formative assessment: 75 practical strategies for linking assessment, instruction, and learning.* Thousand Oaks, CA: Corwin Press and Arlington, VA: NSTA Press.

★ Indicates a strong match between the ideas elicited by the probe and a national standard's learning goal.

National Research Council (NRC). 1996. *National science education standards.* Washington, DC: National Academies Press.

Pugh, K. 2004. Newton's laws beyond the classroom walls. *Science Education* 88 (2): 182–196.

Forces and Newton's Laws

Apple in a Plane

David is sitting in an airplane, flying over the Atlantic Ocean. The plane is moving very fast at a constant speed. He pulls an apple out of his bag and places it on the tray in front of him. Put an X next to all the major forces that are acting on the apple.

____ **A** A force by the tray pushing up on the apple

____ **B** A force by the Earth pulling down on the apple

____ **C** A force by the air pushing down on the apple

____ **D** A force by the air pushing up on the apple

____ **E** A force by the plane in the direction that the plane is moving

____ **F** A force by the apple holding it onto the tray

____ **G** No forces are acting on the apple because it is at rest on the tray.

____ **H** No forces are acting on the apple because it is inside a fast-moving plane.

Explain your thinking. Describe any rules or evidence that you have to support your answer.

_____

_____

_____

_____

_____

_____

_____

Uncovering Student Ideas in Physical Science

Apple in a Plane
Teacher Notes

Purpose
The purpose of this assessment probe is to elicit students' ideas about force related to the interaction between inanimate objects. The probe is designed to determine which forces students think act on an object at rest when it is inside a fast-moving object.

Related Concepts
active action, contact force, gravitational force, interaction, normal force, passive action

Explanation
The best answers are A and B: "A force by the tray pushing up on the apple" (a type of contact force that is called a "normal force" because the force acts perpendicular to the surface) and "A force by the Earth pulling down on the apple" (gravitational force). These forces balance each other so that the motion of the apple is not changing (in this case, the motion of the apple is not changing even though it is inside a fast-moving plane). Some students may answer that the air also exerts a force on the apple. However, because this force is exerted in all directions, cumulatively it is very small and is pointed upward (a buoyancy force) and not downward.

Administering the Probe
This probe is best used with middle school and high school students. Make sure students recognize the apple is sitting on a tray inside a fast-moving plane. It is this feature—an object at rest inside a fast moving object—that distinguishes this probe from the "Apple on a Desk" probe in Keeley, Eberle, and Dorsey (2008).

Related Ideas in *National Science Education Standards* (NRC 1996)

K–4 Position and Motion of Objects
- The position and motion of objects can be changed by pushing or pulling.

5–8 Motions and Forces
- If more than one force acts on an object along a straight line, then the forces will reinforce or cancel one another.

9–12 Motions and Forces
★ Whenever one object exerts a force on another, a force equal in magnitude and opposite in direction is exerted on the first object.
- Gravitation is a universal force that each mass exerts on any other mass.

Related Ideas in *Benchmarks for Science Literacy* (AAAS 1993, 2009)

3–5 Forces of Nature
- The Earth's gravity pulls any object toward it without touching it.

6–8 Forces of Nature
- Every object exerts gravitational force on every other object.

9–12 Motion
★ Whenever one thing exerts a force on another, an equal amount of force is exerted back on it.

Related Research
- Students tend to distinguish between active objects and objects that support, block, or otherwise act passively, such as a table. Students tend to recognize the active actions as forces but often do not consider passive actions to be forces. Teaching students to integrate the concept of passive support into the broader concept of force is challenging, even at the high school level (AAAS 1993).
- Some students believe that if a body is not moving, there is no force acting on it (AAAS 1993). Elementary students typically do not understand gravity as a force. If students do view weight as a force, they often think it is the air that exerts a downward force (AAAS 1993).
- Sjoberg and Lie (1981) found that the state of rest is widely regarded by students as a natural state in which no forces are acting on an object. Furthermore, Minstrell (1982) used a question (which this probe was based on) that asked students to describe the forces acting on a book resting on a table. He found that students had several ideas about the stationary object: gravity kept the book in place; air pressure kept the book in place; the table was "in the way" of the book's falling; an object in contact with the Earth, like a book on the ground, no longer experiences the force of gravity; and a downward force on the book must be greater than an upward force (otherwise, the book would float away). He found the table "being in the way" was the most widely held view (Driver et al. 1994, p. 156).

Suggestions for Instruction and Assessment
- This probe can be combined with the "Apple on a Desk" probe in *Uncovering Student Ideas in Science, Vol. 3* (Keeley, Eberle, and Dorsey 2008). In that probe, the apple is at rest on a table in a house (although the Earth is also moving quite fast) whereas in this probe the apple is on a tray inside a fast-moving object. It may be useful to administer both probes to see if students answer differently depending on whether the apple is on a stationary object or in an airplane moving at a constant, fast speed.
- Provide students with a sequence of scenarios that demonstrate that all surfaces deform in a springlike fashion when objects are placed on them and that the tendency of surfaces to return to their original shape

---

★ Indicates a strong match between the ideas elicited by the probe and a national standard's learning goal.

causes them to exert force on the object. This type of bridging analogy is especially effective with high school students (Clement 1993; Minstrell 1982).

- It is difficult to convince some students that any forces at all are acting in the situation given in the probe. Ask them, *What would happen if you took the tray away?* They should reply that the apple would fall. To "feel" that a force is exerted by the tray on the apple to keep the apple at rest, ask students to hold out one hand, palm flat and upward, and pretend the hand is the tray. Then put a heavy object in the palms of their hands and ask them to hold their hands so the object is not moving. They will notice how they must continually push upward on the object in order for it not to move. Help them see how this force balances the force of gravity, which would cause the object to fall if they removed their hands. If they push harder on the object or relax their muscles, the forces would be unbalanced and the object would move upward or downward.
- It is often difficult for students of all ages (K–12) to grasp that force is an interaction between a pair of objects. To help them internalize this concept, encourage them to identify all the forces in the probe situation as *interactions* instead of just naming the forces. For example, instead of naming gravity as one of the forces, have them describe the interaction as "gravitational force by the Earth on the apple."

References

American Association for the Advancement of Science (AAAS). 1993. *Benchmarks for science literacy.* New York: Oxford University Press.

American Association for the Advancement of Science (AAAS). 2009. Benchmarks for science literacy online. *www.project2061.org/publications/bsl/online*

Clement, J. 1993. Using bridging analogies and anchoring intuitions to deal with students' preconceptions in physics. *Journal of Research in Science Teaching* 30 (1): 1241–1257.

Driver, R., A. Squires, P. Rushworth, and V. Wood-Robinson. 1994. *Making sense of secondary science: Research into children's ideas.* London: RoutledgeFalmer.

Keeley, P., F. Eberle, and C. Dorsey. 2008. *Uncovering student ideas in science: Another 25 formative assessment probes, vol. 3.* Arlington, VA: NSTA Press.

Minstrell, J. 1982. Explaining the "at rest" condition of an object. *The Physics Teacher* 20: 10–14.

National Research Council (NRC). 1996. *National science education standards.* Washington, DC: National Academies Press.

Sjoberg, S., and S. Lie. 1981. *Ideas about force and movement among Norwegian pupils and students.* Institute of Physics Report Series: Report 81-11. Oslo, Norway: University of Oslo.

Forces and Newton's Laws

Ball on a String

Philippe is playing a game at the county fair. He is trying to hit a target—a long pole—by twirling a ball on a string. The ball is making a circular motion over Philippe's head (**see the diagram below**). The arrow shows the direction the ball is twirling in. When do you think that Philippe should let go of the string in order to hit the target? Circle your prediction (A, B, C, D, or E) on the diagram below.

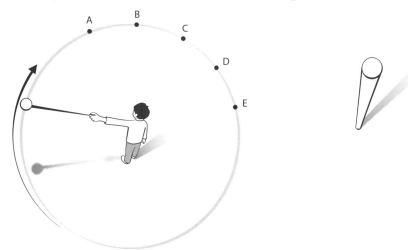

Explain your thinking. On the diagram, draw the ball's path after it leaves the point marked by the letter that you circled.

_____
_____
_____
_____

Uncovering Student Ideas in Physical Science

Forces and Newton's Laws

Ball on a String

Teacher Notes

Purpose

The purpose of this assessment probe is to elicit students' ideas about Newton's first law in the context of circular motion. The probe is designed to determine whether students recognize that an object will move in a straight line unless acted on by an outside force.

Related Concepts

centripetal force, circular motion, Newton's first law

Explanation

The best answer is C: The ball will move in a straight line directly toward the target after it is let go. If no outside forces act on an object, an object will continue to travel in a straight line at a constant speed. As the ball twirls on the end of the string around the boy's head, the sideways force (caused by the tension in the string) points toward the center of the circle. This type of force is called a centripetal force because this force keeps the ball moving in a circular path. If this force no longer acts on the object, then the object will no longer travel in a circle around the center. When the boy lets go of the string at point C, this force goes away and the ball will continue to move in a straight line according to Newton's first law. (Newton's first law states that an object will remain at rest or in uniform motion in a straight line unless acted on by an external force.) Because the external force is no longer exerted by the string pulling on the ball, the ball will fly off in a straight-line path toward the target. (*Note:* This straight line is tangent to the circle at point C. This is not a horizontal line, as it would be at point B.)

Some students may confuse "straight line" with "horizontal line" and choose point E. Older students may have seen a different form of this question on the Force Concept Inventory (Halloun and Hestenes 1985) and inadvertently choose point B because this path would be horizontal. However, this path would not hit the target. The correct path is tangent to the circle

Forces and Newton's Laws

at the point of release. This is because the velocity of the ball at any instant as it travels around the circle is also tangent to the circle.

Administering the Probe

This probe can be used with elementary, middle school, and high school students. With younger students, the emphasis should be on describing motion and on making and testing a prediction. The explanation of Newton's law and centripetal force should wait until middle school. Consider demonstrating the context of this probe to your class by tying a soft object to the end of a string and twirling it over your head to demonstrate its circular motion (do not let go as you want students to make a prediction about where the object will land after you let go).

Related Ideas in *National Science Education Standards* (NRC 1996)

K–4 Position and Motion of Objects
- The position and motion of objects can be changed by pushing or pulling.

5–8 Motions and Forces
- ★ The motion of an object can be described by the object's position, direction, and speed.
- ★ Unbalanced forces will cause changes in the speed or direction of an object's motion.

9–12 Motions and Forces
- ★ Objects change their motion only when a net force is applied.

Related Ideas in *Benchmarks for Science Literacy* (AAAS 1993, 2009)

K–2 Motion
- Things move in many different ways, such as straight, zigzag, round and round, back and forth, and fast and slow.

3–5 Motion
- Changes in speed or direction of motion are caused by forces.

6–8 Motion
- ★ An unbalanced force acting on an object changes its speed or direction of motion or both. If the force acts toward a single center, the object's path may curve into an orbit around the center.

9–12 Motion
- ★ Any object maintains a constant speed and direction of motion unless an unbalanced outside force acts on it.

Related Research
- Students' experiences with whirling objects on a string may contribute to their confusion about the direction of the force that the string is exerting on the object. Students may think they are exerting force along the circular path of the object's motion, rather than perpendicular to it, toward the center (Arons 1997).
- Many students think that objects in circular motion are being "thrown outward." They are likely to think this because of the sensation they feel themselves when traveling around curves in vehicles (Arons 1997, p. 121).
- Some students think that when a sideways force is removed, an object continues along a curved path that is less curved than the original. Some students also believe that after the sideways motion (tangential velocity) wears out, the object falls into the center (Roth, Lucas, and McRobbie 2001).

Suggestions for Instruction and Assessment
- This probe can be combined with the "Rolling Marbles" probe in *Uncovering Student Ideas in Science, Vol. 3* (Keeley, Eberle, and

---

★ Indicates a strong match between the ideas elicited by the probe and a national standard's learning goal.

Dorsey 2008). In that probe, a marble is rolling down a curved chute. Students predict what the path of the marble will be as it goes down the chute.

- This probe can be used as a P-E-O-E probe (Keeley 2008). Ask students to *predict* what the motion of the ball will be and *explain* their reasons for their predictions. Students then test their predictions by tying a soft object to the end of a string, swinging it over their heads, letting go, and *observing* the motion of the object (*Note:* The motion can be difficult to view if students twirl the ball very fast). If their observations do not match their predictions, they are encouraged to rethink their predictions and construct *new explanations*. (**Safety Note:** Be sure to use soft objects tied at the end of the string and have students stay clear of the objects when they let go of the string. Use soft nerf balls or a stuffed sock.)
- Encourage students to investigate this phenomenon in different ways. For example, try removing a section of an embroidery hoop and rolling a marble along its inside edge.
- If a playground merry-go-round is available, have a student riding on the merry-go-round release a ball after reaching a certain point. Have other students note the motion of the ball. Does it travel in a straight or curved line?
- Attach a string to the side of a wind-up toy that travels in a straight line. Show students that pulling on the string pulls the toy sideways only. Have them pull on the string and watch the toy travel in a curved path. Then have them release the string and watch the motion of the toy as they do so. After they do this several times in a row, have students compare this slow-motion example and the forces acting on the toy to the motion of the ball on the string.
- If students have had experiences with garden hoses, ask them what direction the water flows in when they turn on a spigot and the water comes out of a coiled hose that is lying flat on the ground. Many students have seen the water shoot out in a straight line rather than a curved path. Because this phenomenon involves a liquid (water) rather than a solid object, it helps them recognize that Newton's first law applies to liquids as well as solids.

References

American Association for the Advancement of Science (AAAS). 1993. *Benchmarks for science literacy.* New York: Oxford University Press.

American Association for the Advancement of Science (AAAS). 2009. Benchmarks for science literacy online. *www.project2061.org/publications/bsl/online*

Arons, A. 1997. *Teaching introductory physics.* New York: John Wiley and Sons.

Halloun, I. A., and D. Hestenes. 1985. Common sense concepts about motion. *American Journal of Physics* (53) 11: 1056–1065.

Keeley, P. 2008. *Science formative assessment: 75 practical strategies for linking assessment, instruction, and learning.* Thousand Oaks, CA: Corwin Press and Arlington, VA: NSTA Press.

Keeley, P., F. Eberle, C. Dorsey. 2008. *Uncovering student ideas in science: Another 25 formative assessment probes, vol. 3.* Arlington, VA: NSTA Press.

National Research Council (NRC). 1996. *National science education standards.* Washington, DC: National Academies Press.

Roth, W., K. Lucas, and C. McRobbie. 2001. Students' talk about rotational motion within and across contexts, and implications for future learning. *International Journal of Science Education* 23 (2): 151–180.

Forces and Newton's Laws

Why Things Fall

Kathy had a heavy metal ball and Rich had a light wooden ball. They dropped the balls from the same height. They were surprised to discover that the two balls reached the ground at the same time. This is what they said after the balls landed:

Kathy: "The force of gravity must be acting the same on both balls."

Rich: "I don't agree. The metal ball weighs more, so the force of gravity is more on the heavier ball."

Whom do you agree with and why?

_____

_____

_____

_____

_____

_____

_____

_____

Uncovering Student Ideas in Physical Science

Forces and Newton's Laws

Why Things Fall

Teacher Notes

Purpose
The purpose of this assessment probe is to elicit students' ideas about falling objects. The probe is designed to find out whether students recognize the role of mass and forces in understanding why heavy and light objects can fall at the same rate.

Related Concepts
acceleration, gravitational force, mass, net force, Newton's second law, weight

Explanation
Rich has the best answer. For a larger mass to accelerate at the same rate as a smaller mass and hit the floor at the same time, the net force acting on the larger mass must be larger. The heavier ball has more mass than the lighter ball. But because the heavier ball's mass is greater, it needs a larger force to make it accelerate at the same rate as the lighter ball. The ratio of the net force to the mass of the ball is the same for both balls, so they fall at the same rate.

There is a stronger gravitational force on the heavy metal ball; however, the stronger pull on this ball is canceled out by the extra effort required to speed it up (accelerate). If yet another ball were dropped that had five times the mass of the metal ball, the force pulling that ball downward would be five times greater, but that ball would also be five times more difficult to speed up. As a result, this even heavier ball would accelerate just as quickly and hit the floor at the same time as the metal ball. (*Note:* This probe assumes that the effect of air resistance on the balls is negligible.)

Administering the Probe
This probe is best used at the upper middle school and high school levels. Consider demonstrating this probe using two balls of different masses (e.g., a wooden ball and a metal ball).

Forces and Newton's Laws

Related Ideas in *National Science Education Standards* (NRC 1996)

5–8 Motions and Forces
- If more than one force acts on an object along a straight line, then the forces will reinforce or cancel one another, depending on their direction and magnitude. Unbalanced forces will cause changes in the speed or direction of an object's motion.

9–12 Motions and Forces
- ★ Objects change their motion only when a net force is applied. Laws of motion are used to calculate precisely the effects of forces on the motion of objects. The magnitude of the change in motion can be calculated using the relationship Force = Mass × Acceleration, which is independent of the nature of the force.

Related Ideas in *Benchmarks for Science Literacy* (AAAS 1993, 2009)

6–8 Motion
- An unbalanced force acting on an object changes its speed or direction of motion, or both.

9–12 Motion
- ★ The change in motion of an object is proportional to the applied force and inversely proportional to the mass.

Related Research
- Students do not always identify a force to account for falling objects. They think objects "just fall naturally" or that the person letting go of the object has caused it to fall (Driver et al. 1994).
- Studies by Osborne (1984) found that students think heavier objects fall faster.
- Students of all ages (including university students) tend to think that heavier objects fall to Earth faster because they have a bigger acceleration due to gravity (Driver et al. 1994).
- A diversity of misconceptions is found at all age levels. Children between the ages of 7 and 9 progress from the idea that things fall because they're not supported to the idea that things fall because they're "heavy." Between the ages of 9 and 13, students begin to use the term *gravity,* an unseen force, to explain falling, and might say "gravity acts just on heavy objects" or "things fall because air is pushing them down." Surprisingly, many high school and college students who can successfully solve numerical problems involving gravity hold qualitative misconceptions similar to those held by much younger students (Kavanaugh and Sneider 2007).
- Champagne, Klopfer, and Anderson (1980) administered the Demonstration, Observation, and Explanation of Motion Test to 110 students enrolled in an introductory physics course at a major U.S. university. In this test, the instructor carries out demonstrations and asks students to observe (and sometimes predict) the motion of objects and then answer questions about what they see. Although 70% of the subjects had studied high school physics, some for two years, the test revealed that four out of five students believed that, all other things being equal, heavier objects fall faster than lighter ones. The researchers also reported that many students believed that objects fall at constant speed, arguing that speed depends only on weight (mass), which also remains constant. Students who had learned that objects accelerate in free fall reconciled their inconsistent beliefs by saying that acceleration must be due to an increasing force of gravity as the object

---

★ Indicates a strong match between the ideas elicited by the probe and a national standard's learning goal.

gets closer to the ground, and they further supported this idea by claiming that there is no gravity in space (Kavanaugh and Sneider 2007).

Suggestions for Instruction and Assessment

- This probe can be combined with a related probe, "Dropping Balls," found in *Uncovering Student Ideas in Science, Vol. 3* (Keeley, Eberle, and Dorsey 2008).
- The fact that heavy objects are more strongly attracted by gravity than are lighter objects but both fall at the same rate seems like a contradiction to many students. Teachers can help students resolve this apparent contradiction by exploring the concept of inertia (that the heavier object requires a greater force to make it move).
- Students will often conclude that "gravitational force acts the same on all objects" when they observe that objects fall at the same rate. When a class is performing these types of free-fall experiments, teachers should remind students that heavier objects experience a larger gravitational force than lighter objects.

References

American Association for the Advancement of Science (AAAS). 1993. *Benchmarks for science literacy.* New York: Oxford University Press.

American Association for the Advancement of Science (AAAS). 2009. Benchmarks for science literacy online. *www.project2061.org/publications/bsl/online*

Champagne, A., L. Klopfer, and J. Anderson. 1980. Factors influencing the learning of classical mechanics. *American Journal of Physics* 48 (12): 10–74.

Driver, R., A. Squires, P. Rushworth, and V. Wood-Robinson. 1994. *Making sense of secondary science: Research into children's ideas.* London: RoutledgeFalmer.

Kavanaugh, C., and C. Sneider. 2007. Learning about gravity I. Free fall: A guide for teachers and curriculum developers. *Astronomy Education Review* 5 (2). http://dx.doi.org/10.3847/AER2006018

Keeley, P., F. Eberle, and C. Dorsey. 2008. *Uncovering student ideas in science, vol. 3: Another 25 formative assessment probes.* Arlington, VA: NSTA Press.

National Research Council (NRC). 1996. *National science education standards.* Washington, DC: National Academies Press.

Osborne, R. 1984. Children's dynamics. *The Physics Teacher* 22 (8): 504–508.

Forces and Newton's Laws

Pulling on a Spool

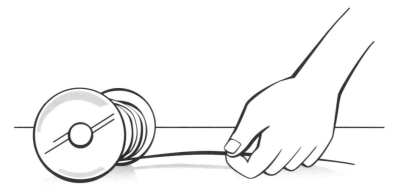

Newton's second law of motion states that the net force acting on an object will cause the object to accelerate in the direction of the net force.

Suriya wants to use wooden spools to make a toy. She first has to figure out how to make a spool roll. She wraps a string around a spool as shown. When Suriya pulls the string slowly to the right, what do you think will happen to the spool? Circle the answer that you think is the best.

A The spool will accelerate to the right in the direction of the pull.

B The spool will accelerate to the left, opposite to the direction of the pull.

C The spool will accelerate neither to the right or left—it will move straight ahead.

Explain the reason for your answer.

___
___
___
___
___
___
___
___

Forces and Newton's Laws

Pulling on a Spool

Teacher Notes

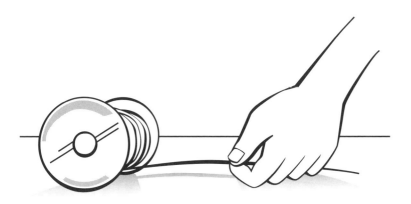

Purpose
The purpose of this assessment probe is to elicit students' ideas about motion in the direction of a net force, using a discrepant event. The probe is designed to determine whether students revert to strongly held intuitive ideas, even when they are presented with a statement of Newton's second law.

Related Concepts
net force, Newton's second law, rolling friction, tension

Explanation
The best response is A: The spool will accelerate to the right in the direction of the pull. If pulled gently (so that the spool does not slide on the surface), the spool will move in the direction of the pull (toward the right). For the spool to roll in that direction, two conditions must be met: (1) the net force must be in the direction of the pull, and (2) the net torque must cause the spool to rotate in the clockwise direction. (Torque is sometimes called the "turning effect" and is the product of the force times the distance between the force and the axis of rotation.)

The primary source of conflict for students is that although the string is pulling to the right, the torque created by the string acting on the spool would create a rotation to the left. However, the force of friction on the spool creates a torque in the opposite direction so that the spool will rotate in the clockwise direction. However, the frictional force can still be less than the force by the string pulling on the spool to the right. Therefore the net force is still to the right.

Administering the Probe
This probe is best used at the high school level. Consider creating a prop—a large spool with a rope attached—to demonstrate this probe. If you use this probe for prediction and discussion, be sure to explicitly state Newton's second law for students before giving them the prompt.

Forces and Newton's Laws

Related Ideas in *National Science Education Standards* (NRC 1996)

9–12 Motions and Forces
★ Objects change their motion only when a net force is applied.
- Laws of motion are used to calculate precisely the effects of forces on the motion of objects.

Related Ideas in *Benchmarks for Science Literacy* (AAAS 1993, 2009)

6–8 Motion
- An unbalanced force acting on an object changes its speed or direction of motion, or both.

9–12 Motion
- Any object maintains a constant speed and direction of motion unless an unbalanced outside force acts on it.

Related Research
- Students often develop strongly held beliefs about the relationship between force and motion based on their everyday experiences. Because of the existence of "hidden forces," such as friction, these everyday experiences often mislead them into constructing a view that is in conflict with a Newtonian view of forces (Champagne, Klopfer, and Anderson 1980).
- Based on their everyday experiences in an environment with friction, many students have a deep-seated belief that "motion implies force." This belief can hinder their understanding of Newton's second law, which describes the relationship between force and acceleration (or a *change* in motion) (Clement 1982).
- Researchers who study conceptual change believe that students need to experience discrepant events in order to change their (incorrect) beliefs (Posner et al. 1982).

Suggestions for Instruction and Assessment
- This probe can be used as a P-E-O-E probe by having each students commit to a *prediction,* provide an *explanation* for his or prediction, and make *observations* to see whether the observations match the prediction. If they don't match, the teacher should encourage students to reconsider and revise their *explanations* (Keeley 2008).
- This probe can be demonstrated with a string wrapped around a spool. The spool must be heavy enough so that the spool does not slide when gently pulled with the string.

References
American Association for the Advancement of Science (AAAS). 1993. *Benchmarks for science literacy.* New York: Oxford University Press.

American Association for the Advancement of Science (AAAS). 2009. Benchmarks for science literacy online. *www.project2061.org/publications/bsl/online*

Champagne, A. B., L. E. Klopfer, and J. H. Anderson. 1980. Factors influencing the learning of classical mechanics. *American Journal of Physics* 48: 10–74.

Clement, J. 1982. Students' preconceptions in introductory mechanics. *American Journal of Physics* 50 (1): 66–71.

Keeley, P. 2008. *Science formative assessment: 75 practical strategies for linking assessment, instruction, and learning.* Thousand Oaks, CA: Corwin Press and Arlington, VA: NSTA Press.

National Research Council (NRC). 1996. *National science education standards.* Washington, DC: National Academies Press.

★ Indicates a strong match between the ideas elicited by the probe and a national standard's learning goal.

Posner, G., K. Strike, P. Hewson, and W. Gertzog. 1982. Accommodation of a scientific conception: Toward a theory of conceptual change. *Science Education* 66 (2): 211–227.

Forces and Newton's Laws

Lifting Buckets

Seth needs to lift a heavy bucket of sand to the second-floor balcony of his house. He ties a rope to the bucket and then stands on the balcony and pulls the bucket straight up. Once the bucket starts moving, how should Seth pull it to get it to move up at a constant speed? Circle your answer.

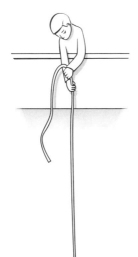

A With a force equal to the weight of the bucket and rope

B With a force greater than the weight of the bucket and rope

C With a force less than the weight of the bucket and rope

Describe your thinking about forces. Provide an explanation to support your ideas.

_____
_____
_____
_____
_____
_____
_____
_____
_____
_____
_____

Uncovering Student Ideas In Physical Science

Lifting Buckets

Teacher Notes

Purpose
The purpose of this probe is to elicit students' ideas about Newton's second law. The probe is designed to determine whether students recognize that in order for an object to move at a constant speed the forces acting on that object must be balanced.

Related Concepts
constant speed, net force, Newton's second law

Explanation
The best answer is A: With a force equal to the weight of the bucket and rope—that is, the pull by the rope must be equal to the weight of the bucket and rope. The key idea is recognizing that the bucket is moving at a constant speed. If the rope were pulling on the bucket (as a result of Seth's pull on the rope) more than the gravitational force by the Earth pulling down, then the bucket would speed up. When an object is moving at constant speed, then the forces acting on the object must be balanced. Many students will likely associate motion (rather than changes in motion) with unbalanced forces.

Administering the Probe
This probe is best used at the high school level. Tell the students that they can consider the mass of the rope to be negligible. They need only compare the force by the person to the weight of the bucket.

Related Ideas in *National Science Education Standards* (NRC 1996)

9–12 Motions and Forces
★ Objects change their motion only when a net force is applied.
• Laws of motion are used to calculate precisely the effects of forces on the motion of objects.

★ Indicates a strong match between the ideas elicited by the probe and a national standard's learning goal.

Forces and Newton's Laws

Related Ideas in *Benchmarks for Science Literacy* (AAAS 1993, 2009)

6–8 Motion
- An unbalanced force acting on an object changes its speed or direction of motion, or both.

9–12 Motion
★ Any object maintains a constant speed and direction of motion unless an unbalanced outside force acts on it.

Related Research
- Many students believe that *motion* implies an unbalanced force rather than that *changes in motion* imply an unbalanced force (Camp and Clement 1994).
- Similar questions have been used in various force-and-motion diagnostic tests. The results from these tests show that the idea that constant motion requires a net force is persistant, even among college physics students after instruction (Halloun and Hestenes 1985).

Suggestions for Instruction and Assessment
- Recent studies have found that some students can benefit from computer simulations of forces and objects.
- One common instructional strategy is to use spring scales to indicate the force that a student applies to an object. Students can then experiment with various situations (such as the situation in this probe) and can measure these forces. Some teachers use electronic "force probes," where the size of the force reading is shown on a computer.

References
American Association for the Advancement of Science (AAAS). 1993. *Benchmarks for science literacy.* New York: Oxford University Press.

American Association for the Advancement of Science (AAAS). 2009. Benchmarks for science literacy online. *www.project2061.org/publications/bsl/online*

Camp, C., and J. Clement. 1994. *Preconceptions in mechanics: Lessons dealing with students' conceptual difficulties.* Dubuque, IA: Kendall Hunt.

Driver, R., A. Squires, P. Rushworth, and V. Wood-Robinson. 1994. *Making sense of secondary science: Research into children's ideas.* London: RoutledgeFalmer.

Halloun, I. A., and D. Hestenes. 1985. Common sense concepts about motion. *American Journal of Physics* 53 (11): 1056–1065.

National Research Council (NRC). 1996. *National science education standards.* Washington, DC: National Academies Press.

★ Indicates a strong match between the ideas elicited by the probe and a national standard's learning goal.

Forces and Newton's Laws

Finger Strength Contest

Max and Joey are having a strength contest (**see illustration at right**). They are using two identical rubber bands to test how much of the rubber band each of them can pull with one finger. They each slip one end of a rubber band around a pole and pull as hard as they can. Joey is able to stretch the rubber band twice as much as Max.

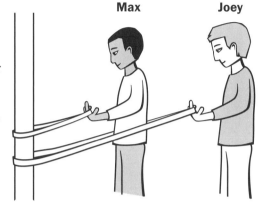

Next, the boys tie two new identical rubber bands together (**see illustration at right**). On the count of three, Max and Joey both pull in opposite directions as hard as they can.

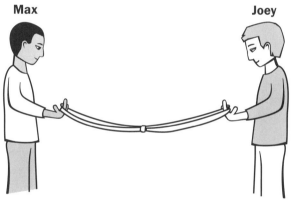

Max's rubber band stretches 16 cm. How far do you think Joey's rubber band will stretch? Circle your answer.

A 32 cm

B 16 cm

C 8 cm

Explain your thinking. What reasoning did you use to decide how far Joey's rubber band will stretch?

_____
_____
_____
_____
_____
_____

Finger Strength Contest

Teacher Notes

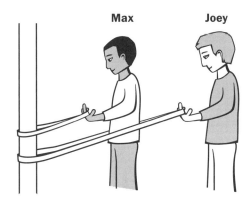

Purpose
The purpose of this assessment probe is to elicit students' ideas about forces. The probe is designed to reveal whether students recognize a situation in which the force applied by one object on another is equal, but opposite, to the force applied by a second object back on the first.

Related Concepts
interaction, Newton's third law

Explanation
The best answer is B: 16 cm. Joey's rubber band will stretch the same amount as Max's rubber band. This is consistent with Newton's third law: The force applied by one object on another is always equal but opposite to the force applied by the second object back on the first. Force is an **interaction** between two objects and is not a property of a single object. The rubber bands serve as a visual reminder that the forces between objects are always equal, but opposite. Joey can only pull on the rubber band as hard as Max pulls.

Administering the Probe
This probe is best used with upper middle school and high school students. Consider demonstrating the probe with real rubber bands before administering it. Demonstrate both activities shown on page 127—that is, (A) each boy pulling on a pole (you might use a table leg) and (B) the two boys pulling in opposite directions.

Related Ideas in *National Science Education Standards* (NRC 1996)

5–8 Motions and Forces
- The motion of an object can be described by its position, direction of motion, and speed.

Forces and Newton's Laws

9–12 Motions and Forces
★ Whenever one object exerts force on another, a force equal in magnitude and opposite in direction is exerted on the first object.

Related Ideas in *Benchmarks for Science Literacy* (AAAS 1993, 2009)

9–12 Motion
★ Whenever one thing exerts a force on another, an equal amount of force is exerted back on it.

Related Research
- Research studies show that students tend to think of a force as a property of single objects rather than as a relationship between two objects. A paired "reaction force" is generally not recognized (Driver et al. 1994).
- A study by Brown (1989) showed that high school students enter physics courses with erroneous preconceptions of Newton's third law. These preconceptions are persistent, and they are difficult to overcome with traditional techniques alone. "The data from the study support the hypothesis that the persistence of preconceptions concerning the third law may result from students' general naive view of force as a property of single objects [rather] than as a relationship between two objects" (Brown 1989, p. 357).
- Some students are apt to use the intuitive rule "more A, more B." Because one boy stretches the single rubber band more than the other boy (as in the top illustration on p. 127), when the boys are on opposite ends of two rubber bands that are tied together (the bottom illustration on p. 127), students are apt to think the boy who stretched the single rubber band more will also stretch the tied-together rubber bands more (Stavy and Tirosch 2000).
- Even teachers can have difficulty in understanding Newton's third law because of the way it was taught to them in traditional classrooms. Action-reaction pairs are often misunderstood as referring to forces acting on a single object (Hughes 2002).

Suggestions for Instruction and Assessment
- Students can easily test their ideas by using identical rubber bands or spring scales. Students should not pull as hard as they can because the rubber bands will likely break or the spring scales may be overloaded. Ask students to try to pull each rubber band (when connected end to end) by a different amount and then observe what happens.
- One way to talk about Newton's third law is to refer to it as the "symmetry principle." According to Bob Prigo, a physicist at Middlebury College, "all forces share a beautiful symmetry property–all forces come in pairs. There is no such thing as a single force. What's more, the pairs are of equal strength and are always oppositely directed. It is important to note, however, that these force pairs are always exerted on different objects" (Prigo 2007, p. 14).
- Equal forces can be illustrated by pushing off of bathroom scales. Take two identical bathroom scales and have two students hold the scales vertically out in front of themselves and against each other so the scales are zeroed. Have the pair of students push on each other, scale to scale. When all motion between the partners has stopped, read the separate scale readings and compare. Repeat by pushing harder. Repeat by pushing more lightly. The "forces," as read off the scales, should be identical in all cases (Prigo 2007).
- Many physics textbooks treat Newton's third law of motion only in passing, often as an addendum to the section covering the con-

★ Indicates a strong match between the ideas elicited by the probe and a national standard's learning goal.

servation of momentum. Teachers should be aware that most students confuse the forces described in the third law with momentum and tend to view force as a property of single objects rather than as a relationship between two objects (Roach 1992).

- Consider rewording Newton's third law of motion for your students (Roach 1992). Many textbooks state that Newton's third law is that "for every action there is an equal and opposite reaction." A better way to state Newton's third law is to say, "When one object exerts a force on a second object, the second object exerts a force on the first that is equal in size and opposite in direction" (McLaughlin and Thompson 1997, p. 110). This phrasing of the third law makes it very clear that there are two objects involved.

References

American Association for the Advancement of Science (AAAS). 1993. *Benchmarks for science literacy.* New York: Oxford University Press.

American Association for the Advancement of Science (AAAS). 2009. Benchmarks for science literacy online. *www.project2061.org/publications/bsl/online*

Brown, D. 1989. Students' concept of force: The importance of understanding Newton's third law. *Physics Education* 24: 353–358.

Driver, R., A. Squires, P. Rushworth, and V. Wood-Robinson. 1994. *Making sense of secondary science: Research into children's ideas.* London: RoutledgeFalmer.

Hughes, M. 2002. How I misunderstood Newton's third law. *The Physics Teacher* 40: 381.

McLaughlin, C., and M. Thompson. 1997. *Physical science.* New York: Glencoe/McGraw-Hill.

National Research Council (NRC). 1996. *National science education standards.* Washington, DC: National Academies Press.

Prigo, R. 2007. *Making physics fun: Key concepts, classroom activities, and everyday examples, K–8.* Thousand Oaks, CA: Corwin Press.

Roach, L. 1992. Demonstrating Newton's third law: Changing Aristotelian viewpoints. *The Science Teacher* 59: 28–31.

Stavy, R., and D. Tirosch. 2000. *How students (mis-) understand science and mathematics: Intuitive rules.* New York: Teachers College Press.

Forces and Newton's Laws

Equal and Opposite

Newton's third law of motion is often stated as, "For every action there is an equal and opposite reaction." Place an X next to each of the statements where the described forces are examples of Newton's third law:

____ **A** You push on a tree with your hand and the tree pushes back on your hand.

____ **B** A small car tows a large truck and they move at constant speed. The truck pulls back on the car and the car pulls on the truck.

____ **C** A small car tows a large truck and they speed up. The truck pulls back on the car and the car pulls on the truck.

____ **D** The Earth pulls down on you (your weight) and the floor pushes up on you.

____ **E** Billy pushes Johnny and causes Johnny to fall down. Johnny exerts a force on Billy and Billy exerts a force on Johnny.

____ **F** You pull on the door of the classroom and the door opens. Your hand exerts a force on the door and the door exerts a force on your hand.

____ **G** A horse is pulling on a cart and the cart is speeding up. The horse exerts a force on the cart and the cart exerts a force on the horse.

____ **H** You hold a book against the wall. You apply a force to the book and the wall applies a force on the book.

Explain your thinking. Describe the rule or reason you used to decide whether a statement fits Newton's third law.

_____
_____
_____

Equal and Opposite

Teacher Notes

Purpose
The purpose of this assessment probe is to elicit students' ideas about pairs of forces that fit Newton's third law. The probe is specifically designed to reveal whether students can identify third law force pairs as involving different objects. The probe is useful in determining whether students misinterpret the familiar colloquial expression of Newton's third law, "For every action there is an equal and opposite reaction."

Related Concepts
interaction, Newton's third law, normal force

Explanation
Statements in which the described forces are examples of Newton's third law are A, B, C, E, F, and G. All of these statements involve equal and opposite forces between two different objects. Statements that do not reflect Newton's third law are D and H. In these examples, there is an interaction of forces but not an interaction of forces as described in the scenarios between two different objects. Both forces are applied to the same object. For example, statement D, "The Earth pulls down on you (your weight) and the floor pushes up on you," involves a gravitational force and the normal force (the floor pushing up). (*Note:* The normal force on a body is generally associated with the force that the surface of one body exerts on the surface of another body in the absence of any frictional forces between the two surfaces.) Likewise, statement H involves forces acting on a single object, the book. In examples D and H, Newton's second law ($F_{net} = ma$) is required to analyze forces acting on a single object. Newton's third law applies to all of the other examples.

Administering the Probe
This probe is best used with upper middle students and high school students. Instead of a paper and pencil probe, it can be administered as a card sort, with students putting the state-

Forces and Newton's Laws

ments that fit Newton's third law in one pile and those that do not in another. There could also be a "we're not sure" pile. Students describe their reasons for putting the cards into the two different categories. They then come up with a "rule" for deciding whether examples of forces in the probe list fit Newton's third law (Keeley 2008). In addition to having students describe why the examples they chose fit Newton's third law, consider having students explain why those they did not choose do not fit Newton's third law.

Related Ideas in *National Science Education Standards* (NRC 1996)

9–12 Motions and Forces
★ Whenever one object exerts force on another, a force equal in magnitude and opposite in direction is exerted on the first object.

Related Ideas in *Benchmarks for Science Literacy* (AAAS 1993, 2009)

9–12 Motion
★ Whenever one thing exerts a force on another, an equal amount of force is exerted back on it.

Related Research
- Studies have revealed that the use of the word *opposite* in the phrase "equal and opposite reaction" may lead some students to think that there is a reaction force acting on the same object rather than two forces involved in an interaction between two objects (Driver et al. 1994).
- A study by Brown (1989) showed that high school students enter physics courses with erroneous preconceptions of Newton's third law. These preconceptions are persistent, and they are difficult to overcome with traditional techniques alone. "The data from the study supports the hypothesis that the persistence of preconceptions concerning the third law may result from students' general naive view of force as a property of single objects [rather] than as a relationship between two objects" (Brown 1989, p. 357).

Suggestions for Instruction and Assessment
- Newton's third law is a statement about the nature of force as an interaction between two objects. A spring can be used to model this interaction, and students can be led to see that when two people push on opposite ends of a spring, the spring compresses and pushes back. The harder the two people push, the more the spring compresses and the more the spring pushes back. Another exercise is to connect two spring scales end to end. Ask one student to pull with one force at one end of the spring and another student to pull with a different force at the other end of the spring. They will discover that no matter how differently they try to pull, the spring scales will equal out and maintain the same reading.
- Encouraging students to identify the interactions between forces—rather than to identify just the forces—may help them distinguish between Newton's third law examples and Newton's second law examples in the list on page 131. For example, in A, the hand pushes against the tree and the tree pushes back on the hand. The interaction involves equal and opposite forces between two objects—the hand and the tree.
- Consider rewording Newton's third law of motion for your students (Roach 1992, p. 29). Many textbooks state that Newton's third law is that for every action there is

---

★ Indicates a strong match between the ideas elicited by the probe and a national standard's learning goal.

an equal and opposite reaction. A better way to state Newton's third law is to say, "When one object exerts a force on a second object, the second object exerts a force on the first that is equal in size and opposite in direction" (McLaughlin and Thompson 1997, p. 110). This phrasing of the third law makes it very clear that there are two objects involved.

- Even teachers can have difficulty understanding Newton's third law because of the way it was taught to them in traditional classrooms. Action-reaction pairs are often misunderstood as referring to forces acting on a single object (Hughes 2002).

References

American Association for the Advancement of Science (AAAS). 1993. *Benchmarks for science literacy.* New York: Oxford University Press.

American Association for the Advancement of Science (AAAS). 2009. Benchmarks for science literacy online. *www.project2061.org/publications/bsl/online*

Brown, D. 1989. Students' concept of force: The importance of understanding Newton's third law. *Physics Education* 24: 353–358.

Driver, R., A. Squires, P. Rushworth, and V. Wood-Robinson. 1994. *Making sense of secondary science: Research into children's ideas.* London: RoutledgeFalmer.

Hughes, M. 2002. How I misunderstood Newton's third law. *The Physics Teacher* 40: 381.

Keeley, P. 2008. *Science formative assessment: 75 practical strategies for linking assessment, instruction, and learning.* Thousand Oaks, CA: Corwin Press and Arlington, VA: NSTA Press.

McLaughlin, C., and M. Thompson. 1997. *Physical science.* New York: Glencoe/McGraw-Hill.

National Research Council (NRC). 1996. *National science education standards.* Washington, DC: National Academies Press.

Roach, L. 1992. Demonstrating Newton's third law: Changing Aristotelian viewpoints. *The Science Teacher* 59: 28–31.

Forces and Newton's Laws

Riding in a Car

Ina, Rie, Kris, and Roberto were sitting in a car. The car suddenly went around a sharp corner. After the car came out of the turn, they wondered about the forces involved. This is what they said:

Ina: "Wow! Did you feel that force pushing us outward? I was pushed against the passenger door."

Rie: "I don't think we were pushed outward. I think we were pushed inward. Otherwise we wouldn't be turning."

Kris: "I could only feel the force pushing us forward. The force must be in this direction because that is the direction we are moving."

Roberto: "Actually, when we started to turn, I think we slowed down a bit, so I think I felt a push backward."

Whom do you most agree with? _____

Explain your thinking about the direction of force on the passengers when the car went around the sharp corner.

_____
_____
_____
_____
_____
_____

Forces and Newton's Laws

Riding in a Car

Teacher Notes

Purpose
The purpose of this assessment probe is to elicit students' ideas about circular motion and forces. The probe is designed to reveal whether students understand that turning requires a force toward the center of the curve.

Related Concepts
circular motion, Newton's first law

Explanation
Rie has the best answer: "I don't think we were pushed outward. I think we were pushed inward. Otherwise we wouldn't be turning." For a car to turn a corner, there must be a force acting on the passengers toward the inside of the corner. When the car begins to turn a corner, you move outward relative to the car until you bump into a person or the door of the car and feel a push. This sensation of being pushed toward the person or car door arises from the fact that we are sensing our motion relative to the car, and not to the road. Relative to the car, we feel we are pushed outward, but relative to the road, the push is inward.

Administering the Probe
This probe is best used with middle school and high school students. Ask students to recall a time they were riding in a car or bus and turned a sharp corner.

Related Ideas in *National Science Education Standards* (NRC 1996)

K–4 Position and Motion of Objects
- The position and motion of objects can be changed by pushing or pulling. The size of the change is related to the strength of the push or pull.

5–8 Motions and Forces
- ★ An object that is not being subjected to a force will continue to move at a constant speed and in a straight line.

★ Indicates a strong match between the ideas elicited by the probe and a national standard's learning goal.

Related Ideas in *Benchmarks for Science Literacy* (AAAS 1993, 2009)

3–5 Motion
- Changes in speed or direction of motion are caused by forces.

6–8 Motion
- An unbalanced force acting on an object changes its speed or direction of motion, or both.
- ★ If a force acts towards a single center, the object's path may curve into an orbit around the center.

9–12 Motion
- ★ Any object maintains a constant speed and direction of motion unless an unbalanced outside force acts on it.

Related Research
- In situations where there is uniform circular motion, some students think there is a centrifugal force pushing the object out rather than a center-directed centripetal force (Roth, Lucas, and McRobbie 2001).
- Many students think that objects in circular motion are being "thrown outward." This is because of the sensation they feel when they are in vehicles traveling around curves (Arons 1997, p. 121).

Suggestions for Instruction and Assessment
- Students need to experience the forces required to move an object in a circle. One method is to roll a bowling ball in a straight line and then have students use a small hammer to exert forces on the ball in order to make the ball turn a corner.
- Have students consider the motion of a car turning a corner from a vantage point above the car (not moving with the car). From this frame of reference, it can be seen that a person in the car is trying to move in a straight line and it is the car that is pushing on the person. Computer simulations of this situation have been found to be effective with students.

References
American Association for the Advancement of Science (AAAS). 1993. *Benchmarks for science literacy.* New York: Oxford University Press.

American Association for the Advancement of Science (AAAS). 2009. Benchmarks for science literacy online. *www.project2061.org/publications/bsl/online*

Arons, A. 1997. *Teaching introductory physics.* New York: John Wiley and Sons.

National Research Council (NRC). 1996. *National science education standards.* Washington, DC: National Academies Press.

Roth, W., K. Lucas, and C. McRobbie. 2001. Students' talk about rotational motion within and across contexts, and implications for future learning. *International Journal of Science Education* 23 (2): 151–180.

★ Indicates a strong match between the ideas elicited by the probe and a national standard's learning goal.

Section 3
Mass, Weight, Gravity, and Other Topics

- Concept Matrix 140
- Related Curriculum Topic
    Study Guides 141
- Related NSTA Press Books, NSTA
    Journal Articles, and NSTA
    Learning Center Resources 141
- 31 Pizza Dough 143
- 32 What Will Happen to
    the Weight? 149
- 33 Weighing Water 153
- 34 Experiencing Gravity 157
- 35 Apple on the Ground 163
- 36 Free-Falling Objects 167
- 37 Gravity Rocks! 171
- 38 The Tower Drop 177
- 39 Pulley Size 181
- 40 Rescuing Isabelle 185
- 41 Cutting a Log 189
- 42 Balance Beam 193
- 43 Lifting a Rock 197
- 44 The Swinging Pendulum 201
- 45 Bicycle Gears 205

Mass, Weight, Gravity, and Other Topics

Concept Matrix
Probes #31–#45

PROBES	#31 Pizza Dough	#32 What Will Happen to the Weight?	#33 Weighing Water	#34 Experiencing Gravity	#35 Apple on the Ground	#36 Free-Falling Objects	#37 Gravity Rocks!	#38 The Tower Drop	#39 Pulley Size	#40 Rescuing Isabelle	#41 Cutting a Log	#42 Balance Beam	#43 Lifting a Rock	#44 Swinging Pendulum	#45 Bicycle Gears
GRADE LEVEL USE →	3–12	3–12	3–12	3–12	3–12	6–12	6–12	3–8	8–12	9–12	3–12	K–5	6–8	3–8	9–12
RELATED CONCEPTS ↓															
acceleration						X									
balancing											X	X			
buoyant force		X													
center of mass											X				
conservation of mass	X														
energy														X	X
free fall						X									
floating		X													
fulcrum													X		
gears and gear ratio															X
gravitational force		X		X	X	X	X	X							
gravitational potential energy								X							
gravity				X	X	X	X	X							
lever													X		
mass	X										X	X			
mechanical advantage									X	X			X		
Newton's second law						X									
Newton's universal law of gravity				X											
pendulum														X	
periodic motion														X	
potential energy													X		
pressure			X												
pulley									X	X					
simple machine									X	X			X		
spherical earth							X								
tension									X	X					
torque											X				
turning effect											X				
variables														X	
weight	X	X	X								X	X			
work										X			X		X

Related Curriculum Topic Study Guides*

Conservation of Matter
Properties of Matter
Measurement, Observation, and Tools
Earth's Gravity
Gravity in Space
Gravitational Force
Pressure and Buoyancy
Work, Power, and Machines
Laws of Motion

* Guides will be found in Keeley, P. 2005. *Science Curriculum Topic Study: Bridging the Gap Between Standards and Practice.* Thousand Oaks, CA: Corwin Press and Arlington, VA: NSTA Press. Each Curriculum Topic Study Guide shows the reader how to Identify Adult Content Knowledge, Consider Instructional Implications, Identify Concepts and Specific Ideas, Examine Research on Student Learning, Examine Coherency and Articulation, and Clarify State Standards and District Curriculum.

Related NSTA Press Books, NSTA Journal Articles, and NSTA Learning Center Resources

NSTA Press Books

American Association for the Advancement of Science (AAAS). 2001. *Atlas of science literacy.* Vol. 1. (See "Gravity" map, pp. 42–43.) Washington, DC: AAAS.

Eisenkraft, A., and L. Kirkpatrick. 2006. *Quantoons.* Arlington, VA: NSTA Press.

Horton, M. 2009. *Take-home physics: High impact, low-cost labs.* Arlington, VA: NSTA Press.

Keeley, P. 2005. *Science curriculum topic study: Bridging the gap between standards and practice.* Thousand Oaks, CA: Corwin Press and Arlington, VA: NSTA Press.

Keeley, P., F. Eberle, and L. Farrin. *Uncovering student ideas in science: 25 formative assessment probes.* Arlington, VA: NSTA Press. (Especially see "Talking About Gravity" probe, pp. 97–102.)

Robertson, W. 2002. *Energy: Stop faking it! Finally understanding science so you can teach it.* Arlington, VA: NSTA Press.

Robertson, W. 2002. *Force and motion: Stop faking it! Finally understanding science so you can teach it.* Arlington, VA: NSTA Press.

Sneider, C. 2003. Examining students' work. In *Everyday assessment in the science classroom,* ed. J. M. Atkin and J. E. Coffey, 27–40. Arlington, VA: NSTA Press.

NSTA Journal Articles

Bar, V., C. Sneider, and N. Martimbeau. 1997. Is there gravity in space? *Science and Children* (Apr.): 38–39.

Bryan, R., A. Laroder, D. Tippens, M. Emaz, and R. Fox. 2008. Simple machines in the community. *Science and Children* (Mar.): 38–42.

Chessin, D. 2007. Simple machine science centers. *Science and Children* (Feb.): 36–41.

Dotger, S. 2008. Using simple machines to leverage learning. *Science and Children* (Mar): 22–27.

Jarrard, A. 2008. The thinking machine: A physical science project. *Science Scope* (Nov.): 24–28.

King, K. 2007. Intertial mass. *Science Scope* (Dec.): 28–35.

Kraft, S., and L. Poyner. 2004. Was the great pyramid built with simple machines? *Science and Children* (Oct): 46–48.

Nelson, G. 2004. What is gravity? *Science and Children* (Sept.): 22–23.

Robertson, W. 2008. Science 101: Do balances and scales determine an object's weight or mass? *Science and Children* (Mar.): 68–71.

Rosenblatt, L. 2004. Those puzzling pendulums. *The Science Teacher* (Dec.): 38–41.

NSTA Learning Center Resources
NSTA Podcasts:
http://learningcenter.nsta.org/products/podcasts.aspx?lid=hp
Explanations of Work
Force and Distance Relationships in Levers–Part I

Force and Distance Relationships in Levers–Part II
Gravity
Gravitational Forces
Introduction to Simple Machines: The Lever
Law of Gravity
Mass and Weight
Weight
Work Defined

NSTA SciGuides:

*http://learningcenter.nsta.org/products/sciguides.
aspx?lid=hp*
Force and Motion

NSTA SciPacks:

*http://learningcenter.nsta.org/products/scipacks.
aspx?lid=hp*
Force and Motion
Gravity and Orbits

NSTA Science Objects:

*http://learningcenter.nsta.org/products/science_objects.
aspx?lid=hp*
Gravity and Orbits: Universal Gravitation
Gravity and Orbits: Gravitational Force

Mass, Weight, Gravity, and Other Topics

Pizza Dough

Nellie makes pizzas at the local pizza parlor. She starts with a ball of pizza dough. She flattens it by hand. She then tosses the flattened dough until it is stretched into the shape of a large circle.

Part 1: Which best describes what happens to the **weight** of the pizza dough after it is stretched out to make a pizza? Circle the best answer.

    A The weight of the dough increases after it is stretched out.

    B The weight of the dough decreases after it is stretched out.

    C The weight of the dough stays the same after it is stretched out.

Part 2: Which best describes what happens to the **mass** of the pizza dough after it is stretched out to make a pizza? Circle the best answer.

    A The mass of the dough increases after it is stretched out.

    B The mass of the dough decreases after it is stretched out.

    C The mass of the dough stays the same after it is stretched out.

Explain your thinking. What rule or reasoning did you use to decide what happens to the weight and mass of the dough after it is stretched?

_____
_____
_____
_____
_____

Mass, Weight, Gravity, and Other Topics

Pizza Dough

Teacher Notes

Purpose
The purpose of this assessment probe is to elicit students' ideas about weight and mass when a property of an object changes. The probe is specifically designed to determine whether students recognize that although weight and mass are different, both the weight and mass of an object stay the same when the object changes shape. Because many students learn that an object's mass is the same on the Earth and the Moon but its weight differs (the object will weigh less on the Moon than on the Earth), they may apply this rule to contexts other than location, such as change in shape.

Related Concepts
conservation of mass, mass, weight

Explanation
The best answer to each part is C: The weight and mass of the dough stay the same after the dough is stretched out. *Weight* and *mass* are terms that students have difficulty distinguishing between. Sometimes when they learn the difference between weight and mass, particularly in contexts such as comparing weight and mass on the Moon versus on Earth, they may develop a misconception that weight and mass must also be different when a property changes, such as the shape of the pizza dough.

On Earth, the gravitational pull of the Earth on an object is measured as weight. Top-loading or hanging-spring scales are typically used to measure weight by the distance a spring stretches or compresses as a result of the load placed on the scale. Weight depends on where an object is located—the weight of an object on top of a high mountain would be a little less than at sea level because the object is further from Earth's center. Weight also differs when an object is located beyond Earth. The weight of an object on the Moon is less than the weight of the object on Earth because the gravitational attraction between the object and the Moon is less than the gravitational attraction between the object and Earth. Likewise,

Mass, Weight, Gravity, and Other Topics

an object would weigh more on Jupiter than on Earth because of the much larger mass of Jupiter and hence the greater pull by Jupiter on the object. However, regardless of where the object is, if the shape of the object is changed in the same location, its weight is unchanged. (*Note:* The radii of the Moon, Earth, and Jupiter also affect the size of the gravitational force acting on an object. The closer the distance to the planet—the smaller the radius—the larger the gravitational force exerted on an object. However, the difference in mass when comparing the gravitational forces on the Moon, Jupiter, and the Earth is far greater than the effect of distance.)

The gravitational attraction between the Earth and the balled and the stretched-out pizza dough is the same because the Earth pulls on the dough in the same way regardless of shape as long as the nearness to the Earth remains the same. Therefore, the weight remains the same.

Mass is the amount of matter in an object. It is independent of location. In school, mass is typically measured with a balance. To do this, an object is placed in one pan and known masses are placed in the opposite pan until the object and masses are balanced. (*Note:* Weight can be measured with a balance as well.) Regardless of where an object is located, on Earth or beyond Earth, its mass always remains the same. In addition, changing the shape of an object does not change its mass. The ball of pizza dough and the stretched-out pizza dough have the same mass. That's because they both contain the same amount of matter. No matter was lost or gained when the dough was stretched; therefore the mass remains the same.

Administering the Probe

This probe can be used with upper elementary, middle school, and high school students once students have learned to distinguish between weight and mass. The teacher can introduce the probe to the class by working with a ball of actual pizza dough, modeling clay, or other substance that can be shaped into a ball and stretched into a pancake shape.

Related Ideas in *National Science Education Standards* (NRC 1996)

K–4 Properties of Objects and Materials
- Objects have many observable properties, including size, weight, shape, color, temperature, and the ability to react with other substances. Those properties can be measured using tools such as rulers, balances, and thermometers.

Related Ideas in *Benchmarks for Science Literacy* (AAAS 1993, 2009)

K–2 Structure of Matter
- Objects can be described in terms of their properties.

K–2 Manipulation and Observation
- Weigh objects using a scale.

6–8 Manipulation and Observation
- Make accurate measurements of length, volume, weight, elapsed time, rates, and temperature by using appropriate devices.

Related Research
- Researchers have found that children, from an early age, notice how objects differ in how they "press down." This "felt weight" is an early conception of the property of weight (Driver at al. 1994).
- The concept of weight as a pulling-down force and the concept of mass develop slowly. The word *mass* is often associated by students with the phonetically similar word

massive, and thus students often think the mass changes if there is a change in size or volume. Students often compare mass by bulk appearance (Driver et al. 1994).
- The physicist's idea of weight as the force of gravity on an object did not appear to be a firmly held idea in studies of secondary students (Ruggiero et al. 1985).

Suggestions for Instruction and Assessment

- This probe can be used as a P-E-O-E probe (Keeley 2008). Have students *predict* what will happen to the weight before and after changing the shape of the pizza dough or similar material and have them support their predictions with *explanations*. Have them test their predictions using a device that measures weight (e.g., top-loading spring scale). When students' *observations* do not match their predictions, encourage them to revisit and revise their *explanations.* Repeat with mass, using a mass-measuring device (e.g., a balance scale).
- Use caution when explaining the difference between weight and mass based on location. Typically, teachers use the-weight-and-mass-on-Earth versus the-weight-and-mass-on-the-Moon as the phenomenon to explain the difference between weight and mass. However, there is a danger that students may develop a generalization that whenever one thing changes (not just location) weight will be different and mass will always stay the same. Be sure to include examples in which *both* weight and mass do not change.
- If the teacher combines a definition of mass versus weight with a discussion of how their measurement devices differ—for example, how the devices function on Earth versus in other locations such as the Moon—students may be better able to conceptualize the difference between weight and mass.
- Teachers should consider the age, experience, and readiness of their students to determine when it is appropriate to distinguish between weight and mass. "Weight" is sometimes used as a stepping-stone to mass because students can conceptualize "felt weight." Although teachers recognize that *mass* is the correct scientific term to use when referring to the amount of matter an object contains, and that the term *weight* refers to the measurement of gravitational pull on the object, some children just are not developmentally ready to learn the distinction. (There are times, however, when it is fine to use *weight* instead of *mass,* such as when applying conservation reasoning.) The national science standards do not introduce mass until middle school; teachers should therefore refer to their own state or local curricula when deciding whether to use *mass* or *weight* with younger children.
- Teachers should help students be aware that throughout our society, the two terms *weight* and *mass* are frequently used interchangeably. In the supermarket, you will find many products on which the net weight is listed both in English and in metric units—for example, on a bag of Hershey's Extra Dark Chocolate pieces, you will find "NET WT 5.1 oz. (144 g)." The English units are indeed a weight (in this case, ounces), but the metric units are actually a measure of mass (grams). The reason this does not cause any real confusion is that as long as one stays on the surface of the Earth, the ratio of mass to weight is fixed (ignoring extremely small variations in the force of gravity on the surface of the Earth).

References

American Association for the Advancement of Science (AAAS). 1993. *Benchmarks for science literacy.* New York: Oxford University Press.

American Association for the Advancement of Science (AAAS). 2009. Benchmarks for science literacy online. *www.project2061.org/publications/bsl/online*

Driver, R., A. Squires, P. Rushworth, and V. Wood-Robinson. 1994. *Making sense of secondary science: Research into children's ideas.* London: RoutledgeFalmer.

Keeley, P. 2008. *Science formative assessment: 75 practical strategies for linking assessment, instruction, and learning.* Thousand Oaks, CA: Corwin Press and Arlington, VA: NSTA Press.

National Research Council (NRC). 1996. *National science education standards.* Washington, DC: National Academies Press.

Ruggiero, S., A. Cartielli, F. Dupre, and M. Vincentini-Missoni. 1985. Weight, gravity, and air pressure: Mental representations by Italian middle-school pupils. *European Journal of Science Education* 7 (2): 181–194.

What Will Happen to the Weight?

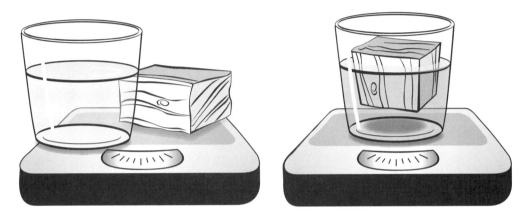

Devon places a wooden block and a bucket of water side by side on a scale. He records the total weight of the objects. Devon then places the wooden block in the bucket so it floats in the water. What do you think will happen to the total weight of the block plus the bucket of water after the wooden block is placed in the bucket of water? Circle your answer.

A The total weight will increase.

B The total weight will decrease.

C The total weight will stay the same.

Explain your thinking. What rule or reasoning did you use to decide what would happen to the total weight?

_____

_____

_____

_____

_____

What Will Happen to the Weight?

Teacher Notes

Purpose
The purpose of the assessment probe is to elicit students' ideas about weight. The probe is designed to determine whether students recognize that the gravitational force on an object, and thus its weight, is the same whether an object is floating in water or is outside of water.

Related Concepts
buoyant force, floating, gravitational force, weight

Explanation
The best answer is C: The total weight will stay the same. Some students answer the question thinking only about the block (and not the system containing both the block and the water). Students who focus only on the block are correct in thinking that the block *appears* to weigh less in the water (due to the upward buoyant force by the water on the block), but the total weight of the system is the same. The force by the water upward on the block (called the buoyant force) is balanced by the force by the block on the water. Therefore, the fact that the block is floating will have no effect on the total weight of the bucket with the water and the block.

Administering the Probe
This probe is best used with upper elementary, middle school, and high school students. Make sure students understand that the block and the bucket of water are side by side on the scale and that the block is then put into the bucket of water while on the scale.

Related Ideas in *National Science Education Standards* (NRC 1996)

K–4 Properties of Objects and Materials
- Objects have many observable properties, including size, weight, shape, color, temperature, and the ability to react with other

substances. Those properties can be measured using tools, such as rulers, balances, and thermometers.

9–12 Motions and Forces
- Gravitation is a universal force that each mass exerts on any other mass.

Related Ideas in *Benchmarks for Science Literacy* (AAAS 1993, 2009)

K–2 Manipulation and Observation
- Weigh objects using a scale.

3–5 Forces of Nature
- The earth's gravity pulls any object on or near the earth toward it without touching it.

6–8 Forces of Nature
- Every object exerts gravitational force on every other object.

6–8 Manipulation and Observation
- Make accurate measurements of length, volume, weight, elapsed time, rates, and temperature by using appropriate devices.

Related Research
- Stead and Osborne (1980) found that 30% of the 13-year-olds they studied thought there was no force of gravity in water and that explains why things float. Other students suggested that there is less gravity in water or even that there is gravity in water but that it acts upward (Driver et al. 1994).
- The physicist's idea of weight as the force of gravity on an object did not appear to be a firmly held idea in studies of secondary students (Ruggiero et al. 1985).
- A study by Watts (1982) found that secondary students have a very flexible view of gravity—namely, that gravity does not act the same way on all objects. In addition, students thought that gravity did not even act the same way at all times on the same object (Driver et al. 1994).

Suggestions for Instruction and Assessment
- This probe can be used as a P-E-O-E activity (Keeley 2008). Provide students with a container of water and a floating object that can be placed in the container. Have students (a) *predict* what the total weight will be before and after adding the object to the container of water, (b) support their predictions with *explanations*, (c) test their predictions, and (d) if their *observations* do not match their predictions, revisit and revise their *explanations*.
- Students' ideas about weight can be probed in different ways, such as by placing an ice cube next to a glass of water and asking students to predict whether or not the total weight will change when the ice is added to the water. Extend the probe to ask what will happen after the ice is all melted.
- It's important to have consistent definitions for *weight* and *gravity* (Morrison 1999).

References
American Association for the Advancement of Science (AAAS). 1993. *Benchmarks for science literacy.* New York: Oxford University Press.

American Association for the Advancement of Science (AAAS). 2009. Benchmarks for science literacy online. *www.project2061.org/publications/bsl/online*

Driver, R., A. Squires, P. Rushworth, and V. Wood-Robinson. 1994. *Making sense of secondary science: Research into children's ideas.* London: RoutledgeFalmer.

Keeley, P. 2008. *Science formative assessment: 75 practical strategies for linking assessment, instruction, and learning.* Thousand Oaks, CA: Corwin Press and Arlington, VA: NSTA Press.

Morrison, R. 1999. Weight and gravity: The need for consistent definitions. *The Physics Teacher* (37): 51.

National Research Council (NRC). 1996. *National science education standards.* Washington, DC: National Academies Press.

Ruggiero, S., A. Cartelli, F. Dupre, and M. Vincentini-Missoni. 1985. Weight, gravity, and air pressure: Mental representations by Italian middle-school pupils. *European Journal of Science Education* 7 (2): 181–194.

Stead, K., and R. Osborne. 1980. Gravity. LISP Working Paper 20. Hamilton, New Zealand: University of Waikato, Science Education Research Unit.

Watts, D. 1982. Gravity: Don't take it for granted! *Physics Education* 17: 116–121.

Mass, Weight, Gravity, and Other Topics

Weighing Water

Container 1   Container 2   Container 3

The same amount of water is poured into three different-shaped containers with identical masses. Each container is placed on a bathroom scale. Circle the observation below that best describes what you would see when you looked at the weight reading on the bathroom scales.

A All three scales will have the same reading.

B Container 1's scale will have the highest reading.

C Container 2's scale will have the highest reading.

D Container 3's scale will have the highest reading.

E Container 1's scale and container 2's scale will have the same reading; container 3's scale will have a different reading.

Explain your thinking. What rule or reasoning did you use to select your answer?

_____
_____
_____
_____
_____
_____
_____
_____

Uncovering Student Ideas in Physical Science

Weighing Water

Teacher Notes

Container 1   Container 2   Container 3

Purpose
The purpose of this assessment probe is to elicit students' ideas about weight. The probe is designed to reveal whether students recognize that the weight of identical volumes of water is the same, regardless of the shape of the container.

Related Concepts
pressure, weight

Explanation
The best response is A: All three scales will have the same reading. The weight in all three containers is the same. Weight is the measure of gravitational force between the Earth and an object. In all three containers, the gravitational force is the same as long as all three containers are in the same location. While the weight stays the same, the pressure differs. Weight and pressure are not the same. When a force is spread out evenly over some surface, pressure is the amount of force exerted for each unit of area (e.g., centimeters^2 or inches^2). Pressure increases when the surface area decreases as long as the exerted force stays the same. In this case, the container with less surface area in contact with the scale exerts more pressure. However, less or more pressure does not change the weight.

Administering the Probe
This probe is best used with upper elementary, middle school, and high school students. Make sure students understand that the amount of water is the same in each container and that each of the containers has the same mass.

Mass, Weight, Gravity, and Other Topics

Related Ideas in *National Science Education Standards* (NRC 1996)

K–4 Properties of Objects and Materials
- Objects have many observable properties, including size, weight, shape, color, temperature, and the ability to react with other substances. Those properties can be measured using tools, such as rulers, balances, and thermometers.

Note: Pressure is not explicitly mentioned in the National Science Education Standards in the context of force and motion. However, it is a concept that is referred to in the context of weather (air pressure). In addition, pressure is related to phenomena that students will encounter in their curriculum materials (e.g., students learn that barometric pressure can be used to predict the weather; high pressure is generally associated with sunny skies and low pressure with storms) and is connected to key ideas about weight and forces that are in the National Science Education Standards.

Related Ideas in *Benchmarks for Science Literacy* (AAAS 1993, 2009)

3–5 Forces of Nature
- The earth's gravity pulls any object on or near the earth toward it without touching it.

Note: Pressure is not explicitly mentioned in the Benchmarks for Science Literacy. However, it is related to phenomena students will encounter in their curriculum materials (e.g., physical science courses often introduce students to Bernoulli's law that uses the law of conservation of energy to predict changes in pressure inside a fluid) and is connected to key ideas about weight and forces that are in the Benchmarks.

Related Research
- Research indicates that ideas about weight and ideas about gravity are separate from each other in the minds of most students (Driver et al. 1994).
- Most available research on pressure is in the context of air or water pressure. This probe is useful in finding out whether students have scientific conceptions of pressure that are different from their conceptions of weight.
- Researchers have found that children, from an early age, notice how objects differ in how they "press down." This "felt weight" is an early conception of the property of weight (Driver at al. 1994).
- Researchers have found that students rarely think of applying Newton's laws when asked questions about fluids or pressure (Heron et al. 2003).

Suggestions for Instruction and Assessment
- Students can further test this idea—that the weight of the same volume of water in containers with identical weights is the same, regardless of the pressure that the container of water exerts—by taking an object and weighing it in different positions. For example, they can weigh a log twice: when it is vertical and when it is lying down on its side. Then they can compare these two weights.
- This task is similar to questions used by noted child psychologist Jean Piaget during interviews with children. A simple version would be to ask students to predict how a scale reading would be different if two objects are sitting side by side or if they are stacked one on top of the other.
- To see if students recognize the effect of surface area on pressure, this probe can be followed up with, or preceded by, a similar probe. Present students with three different-shaped containers (e.g., like those on p. 153) and a piece of foam in place of the bath-

room scale. Ask them to think about what will happen when each container is placed on the piece of foam: A. All three containers will compress the foam the same; B. Container 1 will compress the foam the most; C. Container 2 will compress the foam the most; D. Container 3 will compress the foam the most; or E. Containers 1 and 2 will compress the foam the most, and Container 3 will compress the foam a different amount.

The correct answer is C: Container 2—the container with the smallest surface area touching the foam—will compress the foam the most.

- Ask students why someone can walk on top of deep snow with snowshoes but not with regular shoes. (The snowshoes distribute the weight over a greater surface area so the person does not sink in the snow.)
- An interesting classroom challenge is to ask students to construct a pair of shoes that could be used to walk on eggs (Adair and Loveless 1997).

References

Adair, L., and G. Loveless. 1997. Walking on eggs. *The Physics Teacher* 35: 28.

American Association for the Advancement of Science (AAAS). 1993. *Benchmarks for science literacy*. New York: Oxford University Press.

American Association for the Advancement of Science (AAAS). 2009. Benchmarks for science literacy online. *www.project2061.org/publications/bsl/online*

Driver, R., A. Squires, P. Rushworth, and V. Wood-Robinson. 1994. *Making sense of secondary science: Research into children's ideas*. London: RoutledgeFalmer.

Heron, P., M. Loverude, P. Shaffer, and L. McDermott. 2003. Helping students develop an understanding of Archimedes' principle. II. Development of research-based instructional materials. *American Journal of Physics* 71: 11–88.

National Research Council (NRC). 1996. *National science education standards*. Washington, DC: National Academies Press.

Mass, Weight, Gravity, and Other Topics

Experiencing Gravity

What kinds of objects experience gravitational force? Put an X in the correct boxes.

Object	Yes	No	It depends. (Describe the condition it depends on.)
Ball thrown up in the air			
Rock falling off a cliff			
Rock resting on the ground			
Flying bird			
Bird perched on a branch			
Astronaut on the Moon			
Astronaut in orbit in the Space Shuttle			
Star in outer space			
Fish swimming in water			
Person floating in water			
Stone sinking in water			
Speck of dust			
Speeding car			
Helium balloon floating up in the air			
An object buried in the ground			

What rule or reasoning did you use to decide if an object experiences gravitational force?

Uncovering Student Ideas in Physical Science

Experiencing Gravity

Teacher Notes

Purpose
The purpose of this assessment probe is to elicit students' ideas about gravity. The probe is designed to reveal whether students recognize that gravitational force is universal and that it works on every object in the universe regardless of what the object is doing or where it is.

Related Concepts
gravitational force, gravity, Newton's universal law of gravity

Explanation
The answer is "yes" for all the objects on the list. They all experience gravitational force. Gravity is a universal force of attraction between objects—it affects all matter (mass) anywhere in the universe, whether objects are on or near Earth or far out in outer space. It doesn't matter whether objects on Earth are in air or are sinking or floating in water; the Earth's mass still pulls on them, exerting a gravitational force. Objects in space, where there is no atmosphere, experience gravitational force by other objects in space (some people incorrectly think that air is a prerequisite for gravity). For example, an astronaut on the Moon experiences gravitational force due to the pull toward the Moon. However, the gravitational force on the Moon is 1/6 as strong as the gravitational force on Earth. (A common misconception is that astronauts wear Moon boots to weigh them down and keep them from floating away in space. Actually, astronauts wear the weighted boots so they will feel—in terms of weight—more like they feel on Earth; the boots are not intended to keep the astronauts from floating away.)

Size does not make a difference in whether or not an object experiences gravitational force; however, the mass of both objects does determine the strength of the gravitational force between those objects. The more mass each object has, the stronger the gravitational force is between them. A rock falling off a cliff experiences a greater gravitational force than

Mass, Weight, Gravity, and Other Topics

a speck of dust would experience due to the pull by the Earth on the rock. Objects at rest, such as a rock on the ground, experience gravitational force even though they don't visibly appear to be pulled toward the Earth. Moving objects, such as a ball thrown up in the air, a helium balloon floating upward, or a speeding car, experience gravitational force regardless of the direction or speed of motion.

Administering the Probe

This probe is suitable for all levels, once students are familiar with the concepts of gravity and gravitational force. The national science standards do not connect the idea of force and gravity (gravitational force) until middle school; if you are working with younger students you might revise the probe to ask them which objects experience gravity or are acted on by gravity (younger children associate the word *gravity* with a pull before they develop the idea of gravitational force).

Make sure students understand the directions for the probe. They should consider each object and decide whether or not it experiences gravitational force. In some cases, students may think that there need to be special conditions for this to happen. In that case, they should check off "It depends" and describe the condition "it depends" on. The last part of the probe asks students to provide a rule or reason to explain what types of objects are affected by gravity. It is this part of the probe that will reveal whether students recognize that gravitational force is universal and applies to all objects.

This probe can be used as a paper and pencil task or as a prompt for small-group discussion, followed by a whole-class, sense-making discussion. The probe can also be given as a card sort (Keeley 2008). Write each item in the list on page 157 on a separate card and have students sort the cards into a "experiences gravitational force" pile, a "does not experience gravitational force" pile, or a "it depends/we aren't sure" pile. Make sure students discuss each choice they make, providing justification for why they put a card into a particular pile and settling on a generalized rule or reason to explain that decision.

Related Ideas in *National Science Education Standards* (NRC 1996)

5–8 Earth in the Solar System

- Gravity is the force that keeps planets in orbit around the sun and governs the rest of the motion in the solar system. Gravity alone holds us to the earth's surface and explains the phenomena of the tides.

9–12 Motions and Forces

★ Gravitation is a universal force that each mass exerts on any other mass. The strength of the gravitational attractive force between two masses is proportional to the masses and inversely proportional to the square of the distance between them.

Related Ideas in *Benchmarks for Science Literacy* (AAAS 1993, 2009)

K–2 Forces of Nature

- Things near the earth fall to the ground unless something holds them up.

3–5 Forces of Nature

★ The earth's gravity pulls any object on or near the earth toward it without touching it.

6–8 The Earth

★ Everything on or anywhere near the earth is pulled toward the earth's center by gravitational force.

★ Indicates a strong match between the ideas elicited by the probe and a national standard's learning goal.

Uncovering Student Ideas in Physical Science

6–8 Forces of Nature

★ Every object exerts gravitational force on every other object. The force depends on how much mass the objects have and on how far apart they are. The force is hard to detect unless at least one of the objects has a lot of mass.

9–12 Forces of Nature

- Gravitational force is an attraction between masses. The strength of the force is proportional to the masses and weakens rapidly with increasing distance between them.

Related Research

- The notion that there must be air in order for gravity to act is a widespread misconception (Driver et al. 1994).
- Students have varied ideas about whether gravity affects objects in water. For example, some students think that there is no force of gravity in water and that is why things float; that there is less gravity in water; that there is gravity in water but it acts upward; or that gravity only acts on the parts of the body above the surface of the water (Driver et al. 1994).
- Some students think gravity only applies to objects on Earth (Driver et al. 1994).
- Some students think gravity only applies to heavy things (Driver et al. 1994).
- In a study by Stead and Osborne (1980), some students accounted for birds being able to stay up in the air because gravity is only present at the Earth's surface (Driver at al. 1994).
- Watts and Gilbert (1985) found that some secondary students think that gravity begins to act when an object begins to fall and that it stops acting when the object lands on the ground.
- In Stead and Osborne's (1980) study of 258 middle school students, 44% said there was no gravity on the Moon. The idea that not all planets have gravity was common. Over 75% said there is no gravity in space. The researchers suggested that the science fiction idea of "weightlessness" contributed to this misconception.
- Palmer (2001) interviewed 56 students in grade 6 (11–12 years old) and 56 students in grade 10 (15–16 years old). Students were asked to identify which objects were acted on by gravity in nine different scenarios and later to justify those choices in follow-up interviews. Only 11% of the students in grade 6 and 29% of the students in grade 10 correctly indicated that gravity acted on all the objects. The students who indicated that gravity did not act in some of the situations gave a variety of answers. The most common of these were that (a) gravity acts only on falling objects but not on objects moving upward, (b) gravity does not act on stationary objects, and (c) gravity does not act on objects buried underground (Kavanaugh and Sneider 2007).

Suggestions for Instruction and Assessment

- Combine this probe with the "Talking About Gravity" probe in Volume 1 of the previous series, *Uncovering Student Ideas in Science* (Keeley, Eberle, and Farrin 2005). "Talking About Gravity" is useful in finding out whether students believe air or an atmosphere is necessary for gravity.
- Research findings (as seen in Related Research) show that many students have many misconceptions about gravity. Thus, it is imperative that teachers at all grade levels start every instructional unit on gravity by checking their students' understanding of concepts that were taught at prior grade levels. This can be done by asking students to predict what would happen in various physical situations and to justify their predictions. These activities should be built

---

★ Indicates a strong match between the ideas elicited by the probe and a national standard's learning goal.

into curricula on gravity at all levels. This can be thought of as a "spiral curriculum," with the teacher making very sure that students have reached preceding levels before setting them on the course to the next level (Kavanaugh and Sneider 2007).

- Engage students in applying gravity ideas to real-world contexts. Most students do not have enough opportunities to think about how Newton's laws apply in these contexts. The mathematically rich problems in textbooks sometimes mask students' misconceptions because they can find the right equation and plug in numbers to get the right answer. Science instruction is most successful when students carry with them the insights of science into their own world of everyday lived experience (Kavanaugh and Sneider 2007).
- One instructional strategy described by Camp and Clement (1994) involves students using rubber bands to model the gravitational force between individual masses. Students connect rubber bands between their fingers on different hands to model this force.
- Palmer (2001) proposed that rather than focus exclusively on misconceptions, teachers might help students who have some correct scientific understandings about gravity to expand the variety of contexts to which their understandings apply (Kavanaugh and Sneider 2007).

References

American Association for the Advancement of Science (AAAS). 1993. *Benchmarks for science literacy.* New York: Oxford University Press.

American Association for the Advancement of Science (AAAS). 2009. Benchmarks for science literacy online. *www.project2061.org/publications/bsl/online*

Camp, C., and J. Clement. 1994. *Preconceptions in mechanics: Lessons dealing with students' conceptual difficulties.* Dubuque IA: Kendall/Hunt Publishing.

Driver, R., A. Squires, P. Rushworth, and V. Wood-Robinson. 1994. *Making sense of secondary science: Research into children's ideas.* London: RoutledgeFalmer.

Kavanaugh, C., and C. Sneider. 2007. Learning about gravity I. Free fall: A guide for teachers and curriculum developers. *Astronomy Education Review* 5 (2). *http://dx.doi.org/10.3847/AER2006018*

Keeley, P. 2008. *Science formative assessment: 75 practical strategies for linking assessment, instruction, and learning.* Thousand Oaks, CA: Corwin Press and Arlington, VA: NSTA Press.

Keeley, P., F. Eberle, and L. Farrin. 2005. *Uncovering student ideas in science: 25 formative assessment probes.* Arlington, VA: NSTA Press.

National Research Council (NRC). 1996. *National science education standards.* Washington, DC: National Academies Press.

Palmer, D. 2001. Students' alternative conceptions and scientifically acceptable conceptions about gravity. *The Australian Science Teachers Journal* 33 (7): 691.

Stead, K., and R. Osborne. 1980. *Gravity.* LISP Working Paper 20. Hamilton, New Zealand: University of Waikato, Science Education Research Unit.

Watts, D., and J. Gilbert. 1985. *Appraising the understanding of science concepts: Gravity.* Guildford, UK: University of Surrey, Department of Educational Studies.

Mass, Weight, Gravity, and Other Topics

Apple on the Ground

Mr. Rosenberg's tree is full of apples. Some of the apples fall off the tree and land on the ground. Circle the best description of the gravitational force acting on the apple as it sits on the ground.

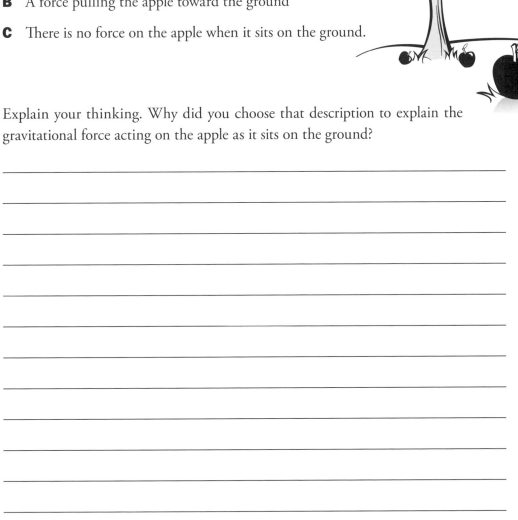

A A force pushing the apple down onto the ground

B A force pulling the apple toward the ground

C There is no force on the apple when it sits on the ground.

Explain your thinking. Why did you choose that description to explain the gravitational force acting on the apple as it sits on the ground?

_____
_____
_____
_____
_____
_____
_____
_____
_____
_____
_____
_____

Uncovering Student Ideas in Physical Science

Apple on the Ground

Teacher Notes

Purpose
The purpose of this assessment probe is to elicit students' ideas about gravity. The probe is designed to reveal whether students recognize that gravitational force is a pull by the Earth, toward the Earth, regardless of whether an object is falling or is stationary.

Related Concepts
gravitational force, gravity

Explanation
The best response is B: A force pulling the apple toward the ground. Gravity is an attractive force that describes the interaction between two objects. In this case it is the interaction between the apple and the Earth in which the Earth exerts a force on the apple that pulls it toward the Earth. It doesn't matter if the apple rests on the ground, hangs on the tree, or falls off the tree. The Earth will pull on the apple in all of these situations. The Earth also pushes up on the apple but this contact force is not a gravitational force. Although we use the language of "pushes" and "pulls" with children, scientists do not differentiate between a push and a pull. However, when teachers use "push" and "pull" with children, it is important to clarify that if two objects are attracted toward each other, then we call this a pulling force; if the two objects are being repelled away from each other, then this would be a pushing force. Using this definition, gravity is always a pulling force.

Administering the Probe
This probe can be used at all levels once students are familiar with the concept of gravity or gravitational force. Because national standards do not connect the ideas of force and gravity (gravitational force) until middle school, you might revise the directions for the probe for younger students by asking them to choose the best description of the gravity acting on the apple on the ground (the word *gravity* is associated with a pull in the younger grades before the idea of gravitational force is developed).

Accompany this change with a change in distracters: Change A to "Gravity pulls the apple toward the ground," change B to "Gravity pushes the apple down onto the ground," and change C to "There is no gravity acting on the apple when it sits on the ground." Make sure the students understand that the apple is resting on the ground and is not moving.

Related Ideas in *National Science Education Standards* (NRC 1996)

5–8 Earth in the Solar System
- Gravity is the force that keeps planets in orbit around the sun and governs the rest of the motion in the solar system. Gravity alone holds us to the earth's surface and explains the phenomena of the tides.

9–12 Motions and Forces
- Gravitation is a universal force that each mass exerts on any other mass. The strength of the gravitational attractive force between two masses is proportional to the masses and inversely proportional to the square of the distance between them.

Related Ideas in *Benchmarks for Science Literacy* (AAAS 1993, 2009)

K–2 Forces of Nature
- Things near the earth fall to the ground unless something holds them up.

3–5 Forces of Nature
★ The earth's gravity pulls any object on or near the earth toward it without touching it.

6–8 The Earth
- Everything on or anywhere near the earth is pulled toward the earth's center by gravitational force.

6–8 Forces of Nature
- Every object exerts gravitational force on every other object. The force depends on how much mass the objects have and on how far apart they are. The force is hard to detect unless at least one of the objects has a lot of mass.

9–12 Forces of Nature
- Gravitational force is an attraction between masses. The strength of the force is proportional to the masses and weakens rapidly with increasing distance between them.

Related Research
- Elementary students typically do not see gravity as a force (Driver et al. 1994).
- Holding, rather than pulling, seems to be a common perception of gravity bound up with the idea that gravity is associated with air pushing down and that an atmosphere of air prevents things from floating away (Driver et al. 1994).
- Watts and Gilbert (1985) found that some secondary students think that gravity begins to act when an object begins to fall and that it stops acting when the object lands on the ground.
- Some manifestations of force are not recognized as pushes and pulls. Some pulls are considered as just holding something in place (Driver et al. 1994).
- Some students believe gravity acts upward. Quite a few high school students and even some college students believe that gravity pushes upward rather than acting to pull things down to Earth's surface (Kavanaugh and Sneider 2007).
- Palmer (2001) interviewed 56 students in grade 6 (11–12 years old) and 56 students in grade 10 (15–16 years old). Students were asked to identify which objects were acted on by gravity in nine different scenarios and later to justify those choices in follow-up

---

★ Indicates a strong match between the ideas elicited by the probe and a national standard's learning goal.

Mass, Weight, Gravity, and Other Topics

interviews. One common response was that gravity does not act on stationary objects (Kavanaugh and Sneider 2007).

Suggestions for Instruction and Assessment

- One reason that gravitational force ideas related to objects at rest are difficult for younger students is that gravity is a force they cannot directly experience and see. Elementary school teachers can cultivate an early conception of gravity at that level by having children engage with phenomena such as magnetism, where unseen forces can be felt and their "pull" effects can be directly observed (Kavanaugh and Sneider 2007).
- Research findings (see Related Research, p. 165) show that many students have misconceptions about gravity. Thus, it is imperative that teachers at all grade levels start every instructional unit on gravity by checking their students' understanding of concepts that were taught at prior grade levels. This can be done by asking students to predict what would happen in various physical situations and to justify their predictions. These activities should be built into curricula on gravity at all levels. This can be thought of as a "spiral curriculum," with the teacher making very sure that students have reached preceding levels before setting them on the course to the next level (Kavanaugh and Sneider 2007).
- Extend this probe by asking students if there a gravitational force acting on the apple when it is attached to the branch of a tree and not moving. Is there a gravitational force when the apple detaches from the branch and is falling?
- One instructional strategy described by Camp and Clement (1994) involves students using rubber bands to model the gravitational force between individual masses. Students connect rubber bands between their fingers on different hands to model this force. When using this activity, make sure that students don't get the idea that the force increases when the distance between the objects increases.

References

American Association for the Advancement of Science (AAAS). 1993. *Benchmarks for science literacy.* New York: Oxford University Press.

American Association for the Advancement of Science (AAAS). 2009. Benchmarks for science literacy online. *www.project2061.org/publications/bsl/online*

Camp, C., and J. Clement. 1994. *Preconceptions in mechanics: Lessons dealing with students' conceptual difficulties.* Dubuque IA: Kendall/Hunt Publishing.

Driver, R., A. Squires, P. Rushworth, and V. Wood-Robinson. 1994. *Making sense of secondary science: Research into children's ideas.* London: RoutledgeFalmer.

Kavanaugh, C., and C. Sneider. 2007. Learning about gravity I. Free fall: A guide for teachers and curriculum developers. *Astronomy Education Review* 5 (2). *http://dx.doi.org/10.3847/AER2006018*

National Research Council (NRC). 1996. *National science education standards.* Washington, DC: National Academies Press.

Watts, D., and J. Gilbert. 1985. *Appraising the understanding of science concepts: Gravity.* Guildford, UK: University of Surrey, Department of Educational Studies.

Mass, Weight, Gravity, and Other Topics

Free-Falling Objects

A group of students was investigating how objects fall. They found five different smooth, spherical objects that had similar shapes but different masses. They numbered the objects in order: Object #1 was the lightest and Object #5 was the heaviest. They then dropped each object from the roof of the school to see how long it would take for each object to reach the ground. The students recorded their results in a table. Circle the table below (A, B, or C) that you think is closest to what happened when the objects were dropped.

Table A

Object Number	Time (seconds)
1	1.6
2	1.8
3	2.0
4	2.2
5	2.4

Table B

Object Number	Time (seconds)
1	2.4
2	2.2
3	2.0
4	1.8
5	1.6

Table C

Object Number	Time (seconds)
1	2.0
2	2.0
3	2.0
4	2.0
5	2.0

Explain your thinking. Describe how the data table supports your ideas about how the objects fall.

_____
_____
_____
_____
_____
_____

Uncovering Student Ideas in Physical Science

Free-Falling Objects

Teacher Notes

Purpose
The purpose of this assessment probe is to elicit students' ideas about how objects with different masses (or weights) behave in free fall where air resistance is negligible. The probe shows whether students can relate their beliefs about how objects fall to a table of numbers.

Related Concepts
acceleration, free fall, gravity, Newton's second law

Explanation
The best answer is Table C: The duration of fall is the same for all five similarly shaped objects when air resistance is negligible. Realistically, there would be some variation in duration of fall given some air resistance, but overall, the time it takes for each object to land on the ground is about the same. Therefore, C is the best answer because the acceleration of all objects in free fall is the same near the surface of the Earth if we assume that there is no air resistance (*Note:* The value of this acceleration is about 9.8 m/s^2). The reason this happens is that the gravitational force is proportional to the mass of an object. However, the acceleration of an object is inversely proportional to its mass. Therefore, heavier objects experience a greater gravitational force (they weigh more). However, because they weigh more, they are more difficult to accelerate, based on Newton's second law. The result is that they will fall at the same rate regardless of their mass.

In this probe, the objects are all released from rest and the distance the objects fall is the same. Therefore, the fact that they take the same amount of time to fall means that they must also have the same acceleration.

Administering the Probe
This probe is best used with middle school and high school students. You might show props to students to give them a context for the probe—for example, five balls with different masses or five other similarly shaped objects of differ-

Mass, Weight, Gravity, and Other Topics

ent masses that would experience minimal air resistance. Modify the tables or consider having students label the 1s in each data table as "lightest" and the 5s as "heaviest" in case they overlook the description in the prompt.

Related Ideas in *National Science Education Standards* (NRC 1996)

5–8 Motions and Forces
- Unbalanced forces will cause changes in the speed or direction of an object's motion.

9–12 Motions and Forces
★ Objects change their motion only when a net force is applied. Laws of motion are used to calculate precisely the effects of forces on the motion of objects. The magnitude of the change in motion can be calculated using the relationship Force = Mass × Acceleration, which is independent of the nature of the force.

Related Ideas in *Benchmarks for Science Literacy* (AAAS 1993, 2009)

3–5 Motion
- Changes in speed or direction of motion are caused by forces. The greater the force is, the greater the change in motion will be.

3–5 Forces of Nature
- The earth's gravity pulls any object toward it without touching it.

6–8 Motion
- An unbalanced force acting on an object changes its speed or direction of motion, or both.

9–12 Motion
★ The change in motion of an object is proportional to the applied force and inversely proportional to the mass.

Related Research

- Students do not always identify a force to account for falling objects. They think objects just "fall naturally" or that the person letting go of the object has caused it to fall (Driver et al. 1994).
- Studies by Osborne (1984) found that students think heavier objects fall faster.
- Students all the way up through university level think that heavier objects fall to Earth faster because they have a bigger acceleration due to gravity (Driver et al. 1994).
- Students might use an intuitive rule—"more A, more B"—to reason that as an object's mass increases, it falls faster (Stavy and Tirosh 2000).
- Champagne, Klopfer, and Anderson (1980) administered the Demonstration, Observation, and Explanation of Motion Test to 110 students enrolled in an introductory physics course at a major U.S. university. In this test, the instructor carries out various demonstrations and asks students to observe (and sometimes predict) the motion of objects and to answer questions about what they see. Although 70% of the subjects had studied high school physics, some for two years, the test revealed that four out of five students believed that, all other things being equal, heavier objects fall faster than lighter ones. Many students also believed that objects fall at constant speed, arguing that speed depends only on weight (mass), which also remains constant. Students who had learned that objects accelerate in free fall reconciled their inconsistent beliefs by saying that

---

★ Indicates a strong match between the ideas elicited by the probe and a national standard's learning goal.

acceleration must be due to an increasing force of gravity as the object gets closer to the ground, and they supported this idea by claiming that there is no gravity in space (Kavanaugh and Sneider 2007).

Suggestions for Instruction and Assessment

- This probe can be combined with the "Dropping Balls" probe in *Uncovering Student Ideas in Science, Vol. 3* (Keeley, Eberle, and Dorsey 2008) or "Why Things Fall" on page 115 in this book.
- This probe can be tested by dropping five similarly shaped objects of different masses. Realize, however, that some effects of air resistance can make it hard to reproduce expected results.
- Connect this probe scenario to the historical example of Galileo's famous experiments with falling objects.
- The idea that heavy objects are more strongly attracted by gravity than lighter objects yet both fall at the same rate seems to be a contradiction for many students. This apparent contradiction can be resolved by discussing the concept of inertia (that the heavier object requires a greater force to make it move).
- Student will often conclude that "gravitational force acts the same on all objects" when they observe that objects fall at the same rate. When performing these types of free-fall experiments, teachers should remind students that heavier objects experience a larger gravitational force than lighter objects.

Caution: It is not recommended that you take students up on a roof to duplicate this scenario. Instead, find a high-enough height to drop the objects from but make sure that it is a safe spot where there is no danger of a person falling.

References

American Association for the Advancement of Science (AAAS). 1993. *Benchmarks for science literacy.* New York: Oxford University Press.

American Association for the Advancement of Science (AAAS). 2009. Benchmarks for science literacy online. *www.project2061.org/publications/bsl/online*

Champagne, A., L. Klopfer, and J. Anderson. 1980. Factors influencing the learning of classical Mechanics. *American Journal of Physics* 48 (12): 10–74.

Driver, R., A. Squires, P. Rushworth, and V. Wood-Robinson. 1994. *Making sense of secondary science: Research into children's ideas.* London: RoutledgeFalmer.

Kavanaugh, C., and C. Sneider. 2007. Learning about gravity I. Free fall: A guide for teachers and curriculum developers. *Astronomy Education Review* 5 (2). http://dx.doi.org/10.3847/AER2006018

Keeley, P., F. Eberle, and C. Dorsey. 2008. *Uncovering student ideas in science, vol. 3: Another 25 formative assessment probes.* Arlington, VA: NSTA Press.

National Research Council (NRC). 1996. *National science education standards.* Washington, DC: National Academies Press.

Osborne, R. 1984. Children's dynamics. *The Physics Teacher* 22 (8): 504–508.

Stavy, R., and D. Tirosch. 2000. *How students (mis-)understand science and mathematics: Intuitive rules.* New York: Teachers College Press.

Mass, Weight, Gravity, and Other Topics

Gravity Rocks!

Three friends were talking about gravity. One friend held up a rock and asked his friends whether the gravitational force on the rock depended on where the rock was located. Each friend had a different idea about a place where the gravitational force on the rock would be the greatest. This is what they said:

Lorenzo: "I think if you put the rock on the top of a very tall mountain, the gravitational force on the rock will be greatest."

Eliza: "I think the gravitational force will be greatest when the rock is resting on the ground near sea level."

Flo: "I think you have to go really high up. If you drop the rock out of a high-flying plane, the gravitational force will be greatest."

Which friend do you most agree with? _____. Explain why you agree with that friend.

_____
_____
_____
_____
_____
_____
_____
_____
_____
_____
_____
_____

Mass, Weight, Gravity, and Other Topics

Gravity Rocks!

Teacher Notes

Purpose
The purpose of this assessment probe is to elicit students' ideas about gravity. The probe is designed to reveal whether students recognize that objects closest to the center of Earth's mass experience the greatest gravitational force. The probe also reveals whether students confuse energy of position (potential energy) with gravitational force.

Related Concepts
gravitational force, gravitational potential energy, gravity

Explanation
The best answer is Eliza's: "I think the gravitational force will be greatest when the rock is resting on the ground near sea level." This is because the rock at sea level is closer to Earth's center of mass than when it is at the top of a tall mountain or falling from a high-flying airplane. Gravitational force on an object increases the closer the object is to the Earth's surface; gravitational force decreases as the distance between the Earth's surface and an object increases. Mathematically, this is expressed as the force of gravity between two objects decreases as the square of the distance between them. In other words, if you double the distance, the gravitational force is reduced by a factor of four (2^2); if you triple the distance, you reduce the force by a factor of nine (3^2). Therefore, the gravitational force on the rock as it is dropped from the plane is the least, followed by the rock on a mountain, then the rock on the ground near sea level.

Some students will select the rock being dropped from the plane because they are probably confusing gravitational potential energy with gravitational force. The farther the rock is from the Earth, the greater potential energy it has. The energy that was used to raise the rock high above the Earth is released as kinetic energy when the rock falls.

Mass, Weight, Gravity, and Other Topics

Administering the Probe

This probe is best used with middle school and high school students. Middle school is the time when students begin to understand terrestrial gravity as a force directed toward Earth's center. Make sure students understand that in the three views of the children in the probe the mass is the same—only the locations differ. The same rock is used in all three positions.

Related Ideas in *National Science Education Standards* (NRC 1996)

5–8 Earth in the Solar System
- Gravity is the force that keeps planets in orbit around the sun and governs the rest of the motion in the solar system. Gravity alone holds us to the earth's surface and explains the phenomena of the tides.

9–12 Motions and Forces
★ Gravitation is a universal force that each mass exerts on any other mass. The strength of the gravitational attractive force between two masses is proportional to the masses and inversely proportional to the square of the distance between them.

Related Ideas in *Benchmarks for Science Literacy* (AAAS 1993, 2009)

K–2 Forces of Nature
- Things near the earth fall to the ground unless something holds them up.

3–5 Forces of Nature
- The earth's gravity pulls any object on or near the earth toward it without touching it.

6–8 The Earth
- Everything on or anywhere near the earth is pulled toward the earth's center by gravitational force.

6–8 Forces of Nature
★ Every object exerts gravitational force on every other object. The force depends on how much mass the objects have and on how far apart they are. The force is hard to detect unless at least one of the objects has a lot of mass.

9–12 Forces of Nature
★ Gravitational force is an attraction between masses. The strength of the force is proportional to the masses and weakens rapidly with increasing distance between them.

Related Research
- Several studies have been conducted about students' ideas related to the way gravity changes with height above the Earth's surface. Many students do hold the physicists' view that gravity decreases with height above the Earth's surface. However, many of the students who hold this view tend to expect a far bigger decrease in the force of gravity with increasing height than what is actually the case (Driver et al. 1994).
- Studies by Stead and Osborne (1980) found that one-third of the 14-year-old students they sampled thought gravity increased with height above the Earth. Students who hold this "higher-stronger" gravity view assume this applies only as long as objects are in Earth's atmosphere.
- A study by Watts (1982) of students age 12–17 found that students believed that gravitational force increased with height. They appeared to confuse gravity with potential energy in assuming a higher force of gravity at greater heights.

★ Indicates a strong match between the ideas elicited by the probe and a national standard's learning goal.

- Some secondary students think that gravity begins to act when an object begins to fall and that it stops acting when the object lands on the ground (Driver et al. 1994).
- Palmer (2001) interviewed 56 students in grade 6 (11–12 years old) and 56 students in grade 10 (15–16 years old). Students were asked to identify which objects were acted on by gravity in nine different scenarios and later to justify those choices in follow-up interviews. Only 11% of the students in grade 6 and 29% of the students in grade 10 correctly indicated that gravity acted on all the objects. One common response was that gravity does not act on objects buried underground (Kavanaugh and Sneider 2007).

Suggestions for Instruction and Assessment

- Combine this probe with the "Talking About Gravity" probe in *Uncovering Student Ideas in Science, Vol. 1: 25 Formative Assessment Probes* (Keeley, Eberle, and Farrin 2005). "Talking About Gravity" is useful in finding out whether students believe air or an atmosphere is necessary for gravity.
- Instruction should begin by checking understanding of prior learning related to gravity. The finding that many students have misconceptions about gravity suggests that teachers at all grade levels should start every instructional unit on gravity by checking their students' understanding of concepts that were taught at prior grade levels. This can be done by asking students to predict what would happen in various physical situations and to justify their predictions. These activities should be built into curricula on gravity at all levels. This can be thought of as a "spiral curriculum," with the teacher making very sure that students have reached preceding levels before setting them on the course to the next level (Kavanaugh and Sneider 2007).
- As a result of his study, Palmer (2001) proposed that rather than focus exclusively on misconceptions, teachers might help students who have some correct scientific understandings about gravity to expand the variety of contexts to which these correct scientific understandings apply (Kavanaugh and Sneider 2007).
- When teaching about gravitational potential energy, make sure students distinguish between the energy interaction and the force interaction.
- For advanced students, consider adding a fourth distractor: "I think if you bury the rock one kilometer below the Earth's surface, the force of gravity on the rock will be the greatest." Research indicates some students believe that gravity does not act on buried objects. Others might believe that the gravitational force increases as an object gets closer to the center of Earth's mass without realizing this does not apply to objects inside the Earth. Inside the Earth, the gravitational field actually decreases linearly with decreasing distance until you get to the center of the Earth where it is zero. Outside the Earth the gravitational field decreases ($1/r^2$). So the maximum field is actually at the surface of the Earth. A famous problem in calculus called the Shell Theorem is used to derive this relationship.

References

American Association for the Advancement of Science (AAAS). 1993. *Benchmarks for science literacy.* New York: Oxford University Press.

American Association for the Advancement of Science (AAAS). 2009. Benchmarks for science literacy online. *www.project2061.org/publications/bsl/online*

Driver, R., A. Squires, P. Rushworth, and V. Wood-Robinson. 1994. *Making sense of secondary science: Research into children's ideas.* London: RoutledgeFalmer.

Kavanaugh, C., and C. Sneider. 2007. Learning about gravity I. Free fall: A guide for teachers and curriculum developers. *Astronomy Education Review* 5 (2). *http://dx.doi.org/10.3847/AER2006018*

Keeley, P., F. Eberle, and L. Farrin. 2005. *Uncovering student ideas in science, vol. 1: 25 formative assessment probes.* Arlington, VA: NSTA Press.

National Research Council (NRC). 1996. *National science education standards.* Washington, DC: National Academies Press.

Palmer, D. 2001. Students' alternative conceptions and scientifically acceptable conceptions about gravity. *The Australian Science Teachers Journal* 33 (7): 691.

Stead, K., and R. Osborne. 1980. *Gravity.* LISP Working Paper 20. Hamilton, New Zealand: University of Waikato, Science Education Research Unit.

Watts, D. 1982. Gravity: Don't take it for granted! *Physics Education* 17: 116–121. (Guilford, UK: University of Surrey.)

Mass, Weight, Gravity, and Other Topics

The Tower Drop

Imagine it is possible to build a very tall tower on Earth's equator. The tower extends far up into Earth's atmosphere. It is higher than the tallest mountain, higher than most planes fly. Now imagine it is possible for a person to stand at the top of this tower and drop a ball.

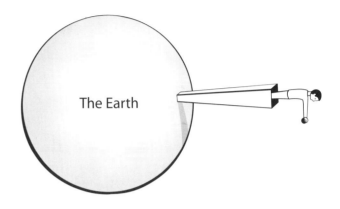

Circle where you think the ball will go when it is dropped:

A The ball will fall to the Earth alongside the tower.

B The ball will fall upward into space, away from Earth.

C The ball will fall downward into space, away from Earth.

D The ball will circle around the Earth.

Draw an arrow from the hand holding the ball to where you think the ball will travel. Your arrow should show the path the ball takes after it is dropped.

Explain your thinking. Why does the ball drop the way your arrow shows?

_____
_____
_____
_____
_____
_____
_____
_____

Mass, Weight, Gravity, and Other Topics

The Tower Drop

Teacher Notes

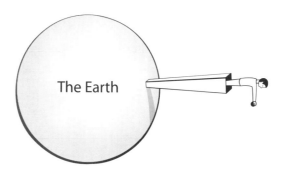

Purpose
The purpose of this assessment probe is to elicit students' ideas about Earth's gravity. The probe is designed as a thought experiment to determine whether students recognize that Earth's gravity pulls falling objects toward its center.

Related Concepts
gravitational force, gravity, spherical Earth

Explanation
The best answer is A: The ball will fall to the Earth alongside the tower. Student drawings should show the ball falling from the hand toward the Earth, about parallel to the tower. All objects fall "down" toward the Earth due to Earth's gravitational pull. Earth's gravity is a center-directed force. "Down" in this context of a spherical Earth means toward Earth's center. Any object dropped on or near Earth (within Earth's gravitational field), including objects dropped from high in Earth's atmosphere, will fall toward Earth's center and eventually stop when they hit the ground. From wherever the object is dropped, a radial line toward the center of the Earth will indicate its path as it falls.

Administering the Probe
This probe is suitable for elementary and middle school students. Emphasize that this is an imaginary tower and that the drawing is not to scale (in reality it would be very difficult for such a tall tower to be straight and rigid). Make sure students know that there are three parts to this probe: (1) the selected response, (2) drawing a line with an arrow to show where the ball falls when dropped from the tower, and (3) a justification for their answer. The probe can be extended by having students draw their ideas on a whiteboard, change the location of the tower to a different location on Earth, and/or make the tower shorter or taller. Note how these changes affect students' responses.

Mass, Weight, Gravity, and Other Topics

Related Ideas in *National Science Education Standards* (NRC 1996)

5–8 Earth in the Solar System
- Gravity is the force that keeps planets in orbit around the sun and governs the rest of the motion in the solar system. Gravity alone holds us to the earth's surface and explains the phenomena of the tides.

9–12 Motions and Forces
- Gravitation is a universal force that each mass exerts on any other mass. The strength of the gravitational attractive force between two masses is proportional to the masses and inversely proportional to the square of the distance between them.

Related Ideas in *Benchmarks for Science Literacy* (AAAS 1993, 2009)

K–2 Forces of Nature
- Things near the earth fall to the ground unless something holds them up.

3–5 Forces of Nature
- ★ The earth's gravity pulls any object on or near the earth toward it without touching it.

6–8 The Earth
- ★ Everything on or anywhere near the earth is pulled toward the earth's center by gravitational force.

6–8 Forces of Nature
- Every object exerts gravitational force on every other object. The force depends on how much mass the objects have and on how far apart they are. The force is hard to detect unless at least one of the objects has a lot of mass.

9–12 Forces of Nature
- Gravitational force is an attraction between masses. The strength of the force is proportional to the masses and weakens rapidly with increasing distance between them.

Related Research
- Some students do not feel the need to identify a force when things fall. They think that things just naturally fall or that someone "let it go" (Driver et al. 1994).
- Studies by Vosniadou (1991) show that students' ideas about the spherical shape of the Earth are closely related to their ideas about gravity and the direction of "down." Students will have difficulty accepting that gravity is center-directed if they do not regard the Earth as spherical (AAAS 1993).
- A study by Nussbaum (1985) gave students a diagram of a very large person standing at the 12:00, 3:00, 6:00, and 9:00 positions along the circumference of the Earth (on opposite sides of the Equator and opposite poles of the Earth). The person is dropping a rock. Students were asked to draw how the rock would fall. Few students drew lines toward the center of the Earth. Many students perceived the direction in which objects fall at different locations on Earth as parallel lines, with an absolute up-and-down dimension, rather than as radial directions pointing toward Earth's center.
- Even though most 15- to 16-year-olds have an Earth-referenced view of "down," a study by Baxter (1989) found that only about 20% had an Earth-centered view rather than an Earth-surface view.

Suggestions for Instruction and Assessment
- Research by Vosniadou (1991) suggests that teachers should teach the concepts of

---
★ Indicates a strong match between the ideas elicited by the probe and a national standard's learning goal.

spherical Earth, space, and gravity in close connection to one another (AAAS 1993).

- Give students a picture of the spherical Earth. Ask them to draw the daytime sky, showing the clouds that are above the Earth. Examine where they draw the clouds. Do students draw clouds on the top of the picture (canopy drawing) or all around the Earth, including over the face of the Earth? Ask further questions to find out more about students' ideas about where the sky is in relation to the Earth. Understanding that the Earth is spherical is necessary in order to understand that objects fall toward the Earth, anywhere on the Earth.

- This probe is based on a similar question in *Earth, Moon, and Stars: Teacher's Guide* (Sneider 1986), published by the GEMS program at the Lawrence Hall of Science. This curriculum guide is an excellent instructional material to develop children's ideas about gravity and the shape of the Earth.

- Combine this probe with different scenarios, such as asking students to describe how an object would fall by moving the location of the tower to the bottom of the drawing (South Pole) and the top of the drawing (North Pole).

- Try a bridging analogy by removing the tower in the illustration on page 178 and drawing a rain cloud over the equator in the 3:00 position, where the tower was. Ask students to draw rain falling from the cloud, showing where the rain goes. If students draw the rain hitting the Earth (but did not say that the ball fell alongside the tower in the original probe), explore their reasons. Connecting this prediction of how rain falls in the revised illustration to how rain falls throughout the world, regardless of where one lives, may help them change their perception of "down" when things above the Earth fall.

References

American Association for the Advancement of Science (AAAS). 1993. *Benchmarks for science literacy.* New York: Oxford University Press.

American Association for the Advancement of Science (AAAS). 2009. Benchmarks for science literacy online. *www.project2061.org/publications/bsl/online*

Baxter, J. 1989. Children's understanding of familiar astronomical events. *International Journal of Science Education* 11 (Special Issue): 502–513.

Driver, R., A. Squires, P. Rushworth, and V. Wood-Robinson. 1994. *Making sense of secondary science: Research into children's ideas.* London: RoutledgeFalmer.

National Research Council (NRC). 1996. *National science education standards.* Washington, DC: National Academies Press.

Nussbaum, J. 1985. The Earth as a cosmic body. In *Children's ideas in science,* ed. R. Driver, E. Guesne, and A. Tiberghien, 170–192. Milton Keynes, UK: Open University Press.

Sneider, C. 1986. *Earth, Moon, and stars: Teacher's guide.* Berkeley, CA: Lawrence Hall of Science.

Vosniadou, S. 1991. Designing curricula for conceptual restructuring: Lessons from the study of knowledge acquisition in astronomy. *Journal of Curriculum Studies* 23: 219–237.

Mass, Weight, Gravity, and Other Topics

Pulley Size

Three friends are discussing the best way to use a pulley to lift a block of wood with a lightweight rope.

Preston: "I think we should use a small pulley. That way we won't have to pull as hard."

Sara: "I disagree. I think if we use a big pulley, it would be easier to lift the wood."

Aliya: "I don't think it matters what size pulley we use. We would have to pull the same."

Which friend do you most agree with? _____ Explain your thinking.

_____
_____
_____
_____
_____
_____
_____
_____

Pulley Size

Teacher Notes

Purpose
The purpose of this assessment probe is to elicit students' ideas about pulleys. The probe is specifically designed to find out whether students think the size of a pulley has an effect on how much easier it is to lift an object (mechanical advantage).

Related Concepts
mechanical advantage, pulleys, simple machine, tension

Explanation
The best answer is Aliya's: "I don't think it matters what size pulley we use. We would have to pull the same." This is because the tension in the rope will be the same at both ends as long as the mass of the rope is small compared to the load. The purpose of a pulley is to change the direction of the pull. Generally, the size of a pulley does not affect the force required to lift a load. Some younger students may predict that the smaller pulley would require less force based on a tendency to use a "less is less" and "more is more" form of reasoning. Older students may predict that the larger pulley will require less force because the point of contact (where the rope pulls on the pulley) is farther from the pivot point (the center of the pulley). Although this is true (creating a longer lever arm), the distance the load is suspended from the pulley is also farther, so that these two effects would cancel each other.

Administering the Probe
This probe is best used with upper middle school and high school students. Make sure students understand that the rope is very light and therefore the mass is not a factor in this problem.

Related Ideas in *National Science Education Standards* (NRC 1996) and *Benchmarks for Science Literacy* (AAAS 1993, 2009)
This probe and the other probes in this section (probes #39–#45) do not explicitly target key ideas in the national standards documents.

Mass, Weight, Gravity, and Other Topics

Both the *National Science Education Standards* (NRC 1996) and *Benchmarks for Science Literacy* (AAAS 1993) deliberately did not include simple machines as important ideas for science literacy. However, because most high school physics courses, many state and local standards, and National Science Foundation–funded reform curricula such as *InterActions in Physical Science; FOSS Simple Machines;* and STC's *Energy, Machines, and Motion* include these concepts as an extension of force, motion, and energy ideas, we decided it was important to include these "Other Topics." In addition, students can use this probe to make a prediction before applying the technological design/engineering process described in the standards to solve design problems without necessarily knowing the specific content related to mechanical advantage, pulleys, and tension.

Related Research

- DiSessa (1993) describes what he calls "phenomenological primitives," or "p-prims," that students use to make simple predictions. A p-prim is a commonsense, intuitive idea used to explain everyday phenomena that often operates below the level of consciousness. One example is the "closer stronger, farther weaker" p-prim that some students may use when asked about lever arms.
- Stavy and Tirosch (2000) describe the "more A, more B" intuitive rule. In this case, students might think that a bigger pulley confers more mechanical advantage (or makes it easier to lift).

Suggestions for Instruction and Assessment

- Once students understand how forces affect the motion of objects, teachers should provide them with experiences that will transfer their learning to thinking about how forces are also involved in the way simple machines work. Make sure they first understand that a simple machine, such as a pulley, is a simple device that affects the force required to perform a given task. This idea later can be connected to the concept of an input force (effort) and an output force (load).
- Several curricula are available for younger students to gain experiences with using pulleys such as the FOSS Levers and Pulley Module and the Science Companion unit on energy. A common instructional strategy is to have students build Rube Goldberg machines—a complex set of mechanical devices (using pullies and levers) built to complete a simple task (Jarrard 2008).

References

American Association for the Advancement of Science (AAAS). 1993. *Benchmarks for science literacy.* New York: Oxford University Press.

American Association for the Advancement of Science (AAAS). 2009. Benchmarks for science literacy online. *www.project2061.org/publications/bsl/online.*

DiSessa, A. 1993. Towards an epistemology of physics. *Cognitive Instruction* 10: 105–225.

Jarrard, A. 2008. The thinking machine: A physical science project. *Science Scope* (Nov.): 24–28.

National Research Council (NRC). 1996. *National science education standards.* Washington, DC: National Academies Press.

Stavy, R., and D. Tirosch. 2000. *How students (mis) understand science and mathematics: Intuitive rules.* New York: Teachers College Press.

Mass, Weight, Gravity, and Other Topics

Rescuing Isabelle

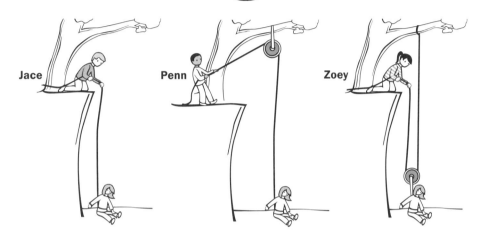

Isabelle hurt her ankle while hiking in the mountains. To get to the road, her friends had to get her up a steep cliff. They have one long rope and one pulley. Her friends discussed the easiest way to pull Isabelle up the cliff. This is what they said:

Jace: "We should tie the rope around Isabelle and pull her up to the road."

Penn: "We should tie the pulley to a tree at the top of the cliff and then pull the rope down through the pulley."

Zoey: "I think we should tie the pulley to Isabelle. Then we should tie one end of the rope to a tree at the top of the cliff and pull the rope up through the pulley."

Which method would require the least force to pull Isabelle up?

____ Jace's method

____ Penn's method

____ Zoey's method

____ It doesn't matter. You would have to pull the same for all three methods.

Explain your thinking. Describe why the method you chose would allow Isabelle's friends to pull her up with the least force.

Uncovering Student Ideas in Physical Science

Rescuing Isabelle

Teacher Notes

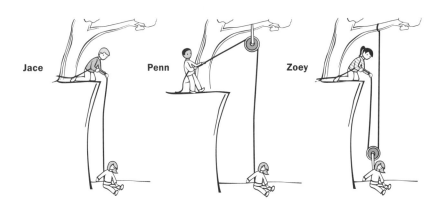

Purpose

The purpose of this assessment probe is to elicit students' ideas related to pulleys and mechanical advantage. The probe is specifically designed to investigate student intuition about *tension*, which is another name for the force exerted by ropes or string.

Related Concepts

mechanical advantage, pulley, simple machine, tension, work

Explanation

The best answer is Zoey's: "I think we should tie the pulley to Isabelle. Then we should tie one end of the rope to a tree at the top of the cliff and pull the rope up through the pulley." Zoey's method would require her friends to pull with a force that is half of Isabelle's weight. Both Jace and Penn's method would require a minimum force that would be equal to Isabelle's weight (double what is required for Zoey's method).

If the friends were to use Penn's method or Jace's method, only one rope would be attached to the load (Isabelle). Therefore, the tension in the rope must be at least equal to Isabelle's weight. In Zoey's method, two ropes are pulling up on Isabelle (one on each side of the pulley). Therefore, the tension in each of these ropes must combine to support Isabelle's weight. The tension in each of these ropes is half of Isabelle's weight. It is interesting to note that the total energy expended to lift Isabelle is the same for all three methods. This is because the force required in Zoey's method is less, but her friends will need to pull twice the length of rope to get her to the top. The work done by Zoey and each of her friends (which is the energy required to lift the load to the top) is the force they apply times the distance they pull the rope. This would be the same for all three methods.

Administering the Probe

This probe is best used with high school students. It can also be used with middle school

students in the context of a technological design/engineering problem. In any case, if possible, have a rope and pulley to show the students.

Related Ideas in *National Science Education Standards* (NRC 1996) and *Benchmarks for Science Literacy* (AAAS 1993, 2009)

This probe and the other probes in this section (probes #39–#45) do not explicitly target key ideas in the national standards documents. Both the *National Science Education Standards* and *Benchmarks for Science Literacy* deliberately did not include simple machines as important ideas for science literacy. However, because most high school physics courses, many state and local standards, and National Science Foundation–funded reform curricula such as *InterActions in Physical Science; FOSS Simple Machines;* and STC's *Energy, Machines, and Motion* include these concepts as an extension of force, motion, and energy ideas, we decided it was important to include these "Other Topics." In addition, students can use this probe to make a prediction before applying the technological design/engineering process described in the standards to solve design problems without necessarily knowing the specific content related to mechanical advantage, pulleys, and tension.

Related Research
- Similar questions have been asked of college physics students. Even after instruction, some of these students experience the same difficulties that we see in younger students (Mazur 1997).

Suggestions for Instruction and Assessment
- A common instructional strategy is to have students build Rube Goldberg machines—a complex set of mechanical devices (using pulley and levers) built to complete a simple task (Jarrard 2008).
- In a Japanese Lesson Study, teachers designed a lesson in which students are challenged to lift heavy loads. The scenario is an earthquake and one of their friends becomes trapped beneath a heavy object. After planning what they might do, they are told by the teacher that they can use levers and pulleys to lift various loads. The title of the task is "Can you lift 100 kilograms?" See *www.lessonresearch.net/canyoulift1.html*.

References

American Association for the Advancement of Science (AAAS). 1993. *Benchmarks for science literacy.* New York: Oxford University Press.

American Association for the Advancement of Science (AAAS). 2009. Benchmarks for science literacy online. *www.project2061.org/publications/bsl/online*

Jarrard, A. 2008. The thinking machine: A physical science project. *Science Scope* (Nov.): 24–28.

Mazur, E. 1997. *Peer instruction: A user's manual.* Upper Saddle River, NJ: Prentice Hall.

National Research Council (NRC). 1996. *National science education standards.* Washington, DC: National Academies Press.

Mass, Weight, Gravity, and Other Topics

Cutting a Log

Hannah, Alex, and Amanda are camping with their family. They decide to go into the forest to find wood for their campfire. They find a large log with no branches on it. One end is very heavy and the other end is light. They decide they need to cut the log in half in order to carry it back to their campsite. They want to make sure that each half is about the same weight. Hannah suggests they balance the log on a rock as shown below and cut the log at the balance point.

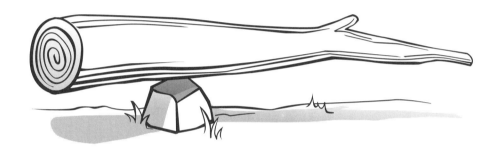

After balancing the log, this is what each said:

Hannah: "I think both pieces will weigh the same."

Alex: "I think the piece on the right side of the balance point will weigh more."

Amanda: "I think the piece on the left side of the balance point will weigh more."

Circle the best answer. Explain your thinking. What rule or reasoning did you use for your answer?

_____
_____
_____
_____
_____
_____

Uncovering Student Ideas in Physical Science   189

Cutting a Log
Teacher Notes

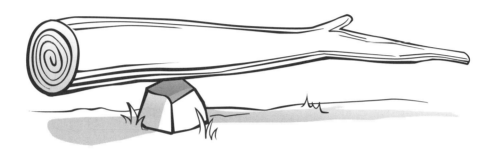

Purpose
The purpose of this assessment probe is to elicit students' ideas about center of mass. The probe is specifically designed to see if students recognize that when an object is balanced on a fulcrum, it does not mean that the parts on either side of the fulcrum will always have equal mass (or weight).

Related Concepts
balancing, center of mass, mass, torque, turning effect, weight

Explanation
The best answer is Amanda's: "I think the piece on the left side of the balance point will weigh more." The shorter piece on the left of the fulcrum (balance point) will have more mass then the longer piece on the right. This is because there is more mass (and more weight) closer to the fulcrum, which has the same effect as less mass (and less weight) farther from the fulcrum. (It is the product of mass times the distance from the fulcrum that must be the same for something to balance.) Many students incorrectly believe that terms such as *center of mass, center of weight,* or *center of gravity* are used to refer to the location where there is the same amount of mass (or weight) on either side of that point. To make things balance, both mass and the location of that mass must be taken into account. For more advanced students, this idea is directly related to the concept of torque, which is sometimes referred to as "turning effect."

Administering the Probe
This probe can be used with elementary, middle school, and high school students. With younger students, the emphasis should be on balancing and comparing the weights on each side, not the more sophisticated concept of center of mass. You can introduce the probe by asking students if they have ever balanced an object where both sides were not the same size. With older students, you may substitute *mass*

Mass, Weight, Gravity, and Other Topics

for *weight* if it doesn't affect students' thinking (some students will confuse the phonetically similar word *massive* with *mass*). Make sure students recognize that the suggestion is to cut the log at the balance point and that once the log is cut, the pieces are two very different lengths. (For younger students, you might point out where the balance point is located.) Make sure that students understand that the weights (or masses) of the two pieces are being compared to each other and that cutting does not change the weight or mass of a piece.

Related Ideas in *National Science Education Standards* (NRC 1996) and *Benchmarks for Science Literacy* (AAAS 1993, 2009)

This probe and the other probes in this section (probes #39–#45) do not explicitly target key ideas in the national standards documents. Both the *National Science Education Standards* and *Benchmarks for Science Literacy* deliberately did not include center of mass as an important idea for science literacy. However, because many elementary curricula include inquiry investigations related to balancing, and middle school curricula often build on these ideas by developing the concept of center of mass, we decided it was important to include these "Other Topics" probes. In addition, students can use this probe to make a prediction before applying the technological design/engineering process described in the standards to solve design problems without necessarily knowing the specific content related to center of mass.

Related Research

- Researchers have investigated children's conceptual development of the meaning of mass and found that this understanding develops slowly. The word *mass* often becomes associated with the phonetically similar word *massive* and in that way is confused with size or volume (Driver et al. 1994).
- From an early age, children have intuitive notions of moments (a *moment* refers to the measure of an object's resistance to changes in its rotational motion) (e.g., Inhelder and Piaget [1958]). When they manipulate a seesaw, for example, they "know" that a "weight" farther away from the center has a bigger effect and they "know" how to achieve a balance using different weights on either side of a beam (Driver et al. 1994, p. 153).
- Some students may use the intuitive rule "more A, more B." Because the right side of the log is longer, they may think there is more wood and thus more mass (or weight) on that side (Stavy and Tirosch 2000).
- Researchers used a similar balancing task (with a baseball bat) with college students. The results showed that even after instruction, many students still believed that the two halves would have equal mass. When probed, some students made statements about balancing, some mentioned torque, and one student said "that's what center of mass means" (Ortiz, Heron, and Shaffer 2005).

Suggestions for Instruction and Assessment

- This probe can be used as a P-E-O-E probe by having students make *predictions* about a similarly shaped object, such as a piece of clay (Keeley 2008). After they make their predictions, students support their predictions with *explanations* and then test their ideas by cutting the object and weighing both pieces. If their *observations* do not match their predictions, they must reconsider their *explanations* in the light of new evidence.

- Connect this probe to a real-life example of a small child and a heavier child trying to balance on a seesaw by having the heavier child move closer to the point of the fulcrum without changing the position of the seesaw on the fulcrum. Would the total weight on each side of the balance point be the same? In this case, the seesaw's position on the fulcrum did not change so the weight of each side of the seesaw, even without the children, is the same. The heavier child would still be heavier than the lighter child. All that changed was the position of the heavier child, which changed the center of mass. Just because they were balanced when the heavier child moved closer to the fulcrum doesn't mean both sides (seesaw plus children) of the fulcrum were of equal weight.
- Pegboard balances can be used so that students can experiment with hanging various masses at different distances from the pivot point. The essential question is "Can you balance two units of mass with one unit of mass?" If students succeed with this task, they can be asked to balance three units with one unit. Eventually students can develop a rule so that they can predict where to hang unequal masses so that they can balance (McDermott 1995).

References

American Association for the Advancement of Science (AAAS). 1993. *Benchmarks for science literacy.* New York: Oxford University Press.

American Association for the Advancement of Science (AAAS). 2009. Benchmarks for science literacy online. *www.project2061.org/publications/bsl/online*

Driver, R., A. Squires, P. Rushworth, and V. Wood-Robinson. 1994. *Making sense of secondary science: Research into children's ideas.* London: RoutledgeFalmer.

Inhelder, B., and J. Piaget. 1958. *The growth in logical thinking from childhood to adolescence.* London: Routledge and Kegan Paul.

Keeley, P. 2008. *Science formative assessment: 75 practical strategies for linking assessment, instruction, and learning.* Thousand Oaks, CA: Corwin Press and Arlington, VA.

McDermott, L. C. 1995. *Physics by inquiry: An introduction to physics and the physical sciences.* Vol. 1. New York: John Wiley and Sons.

National Research Council (NRC). 1996. *National science education standards.* Washington, DC: National Academies Press.

Ortiz, L., P. Heron, and P. Shaffer. 2005. Investigating student understanding of static equilibrium and accounting for balancing. *American Journal of Physics* 73 (6): 545–553.

Stavy, R., and D. Tirosch. 2000. *How students (mis)understand science and mathematics: Intuitive rules.* New York: Teachers College Press.

Mass, Weight, Gravity, and Other Topics

Balance Beam

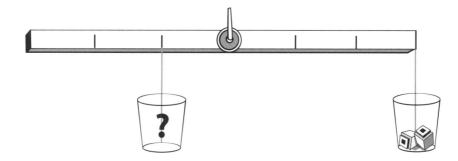

A string is tied around a balance beam and hung so the balance beam is perfectly balanced. One cup is placed at the right end of the beam. Another cup is placed on the left side of the beam. The left cup is closer to the middle than the right cup. Two cubes are placed in the cup on the right. The beam now tips downward on the right side. How many cubes should be placed in the left cup in order to balance the beam? Circle the answer that best matches your thinking.

   2 cubes

   3 cubes

   4 cubes

   5 cubes

   6 cubes

   It is not possible to predict.

Explain your thinking. What rule or reasoning did you use to decide how to get the cups to balance?

_____
_____
_____
_____
_____

Uncovering Student Ideas in Physical Science   193

Mass, Weight, Gravity, and Other Topics

Balance Beam

Teacher Notes

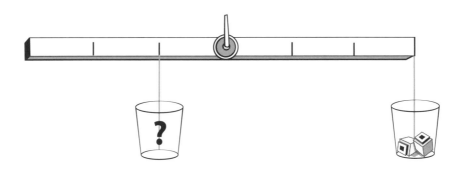

Purpose
The purpose of this assessment probe is to elicit students' ideas about balancing. The probe is specifically designed to see if students recognize that objects with different weights (or masses) can be balanced using a simple mathematical rule.

Related Concepts
balancing, mass, weight

Explanation
The best answer is 6 cubes. This is because masses or weights hanging closer to the pivot have less of an effect on balancing than masses or weights that are hanging farther from the pivot. To create a balance, you must put more weight (or mass) in the left cup.

To find the numerical answer, we need to understand that the distance between the hanging cubes and the pivot point is just as important as the number of cubes. The product of 3 (number of units away from the pivot point) times 2 (the number of cubes) is equal to 6. This number must be equal to the same product on the left side in order for the arm to balance. Because the pan is hanging 1 unit from the pivot point, the number of cubes must be equal to 6 ($1 \times 6 = 6$).

Administering the Probe
This probe is best used with elementary students engaged in inquiry-based activities on balancing and weighing.

Related Ideas in *National Science Education Standards* (NRC 1996) and *Benchmarks for Science Literacy* (AAAS 1993, 2009)
This probe and the other probes in this section (probes #39–#45) do not explicitly target key ideas in the national standards documents. Both the *National Science Education Standards* and *Benchmarks for Science Literacy* deliberately did not include balancing as an important idea

Mass, Weight, Gravity, and Other Topics

for science literacy. However, balancing and weighing are important skills that are developed through inquiry-based activities in the elementary grades that can involve both "messing about" and carefully controlling variables. Balancing is also a way to help young students move beyond their intuitive ideas about balancing and develop mathematical explanations for observations. Because several elementary kit-based science programs such as STC and FOSS include carefully sequenced activities on balancing and weighing, we decided it was important to include this "Other Topics" probe. In addition, students can use this probe to make a prediction before applying the technological design/engineering process described in the standards to solve design problems.

Related Research

- From an early age, children have intuitive notions of moments (a *moment* refers to the measure of an object's resistance to changes in its rotational motion), as identified by researchers such as Inhelder and Piaget (1958). When they manipulate a seesaw, for example, they "know" that a "weight" further away from the center has a bigger effect and they "know" how to achieve a balance using different weights on either side of a beam (Driver et al. 1994, p. 153).

Suggestions for Instruction and Assessment

- Use the P-O-E (predict-observe-explain) strategy with the probe scenario by using a hanging balance beam or a flat board with a fulcrum with interval marks and cups placed at the same interval marks as in the illustration on page 193. Place two uniform objects (such as unifix cubes or pennies) in the cup as in the illustration. Have students commit to *predictions* of how many of the objects they need to put in the other cup to make the beam balance; students test their predictions by making *observations*. If they find their observations do not match their predictions, encourage them to try other numbers of objects until they get the beam to balance. Then ask them to come up with a rule that *explains* how they got the beam to balance. Have them test their rule by putting only one object in the cup on the right (which should balance by putting three objects in the cup on the left).
- Provide students with a balance, fulcrum, and "weights" such as unifix cubes. A balance and fulcrum can be as simple as a flat board and a wedge-shaped block. Ask students to explore different ways to get different numbers of unifix cubes on each side in order to balance the beam. Have them draw and explain the different ways they were able to get their beam to balance.
- Provide younger students with ample opportunities to balance (a) symmetrical and nonsymmetrical objects on a fulcrum and (b) different weights on each side of a fulcrum. From an early age, young children seem to have an intuitive sense that weight and the distribution of weight affect how an object balances even though they may not have the vocabulary to explain the relationship. Listen carefully to their "rules" for balancing and to indications that they are thinking about the relationship between balancing and weight.
- Have students explore and explain the way beam balances work (use both single pan beam balances and double pan beam balances).
- Have students build and balance mobiles to explore the relationship between weight and balance.
- Make a balance beam with a pencil, Popsicle stick, quarter, and a penny. Tape the quarter to one end of the Popsicle stick and the penny to the other end. Lay the stick

across the pencil and find a spot where the stick balances. Have students observe how the lighter coin needs a longer part of the stick to balance with the heavier coin. Ask them what they would have to do differently with the two coins if they were to leave the pencil in the middle of the Popsicle stick. Ask students to explain how this is like two friends of different weights trying to balance on a seesaw.

- Consider preceding this probe with a K-W-L strategy: This is what I *know* about balancing, this is what I *want* to know about balancing, and this is what I *learned* about balancing (Keeley 2008).

References

American Association for the Advancement of Science (AAAS). 1993. *Benchmarks for science literacy.* New York: Oxford University Press.

American Association for the Advancement of Science (AAAS). 2009. Benchmarks for science literacy online. *www.project2061.org/publications/bsl/online*

Driver, R., A. Squires, P. Rushworth, and V. Wood-Robinson. 1994. *Making sense of secondary science: Research into children's ideas.* London: RoutledgeFalmer.

Inhelder, B., and J. Piaget. 1958. *The growth in logical thinking from childhood to adolescence.* London: Routledge and Kegan Paul.

Keeley, P. 2008. *Science formative assessment: 75 practical strategies for linking assessment, instruction, and learning.* Thousand Oaks, CA: Corwin Press and Arlington, VA: NSTA Press.

National Research Council (NRC). 1996. *National science education standards.* Washington, DC: National Academies Press.

Mass, Weight, Gravity, and Other Topics

Lifting a Rock

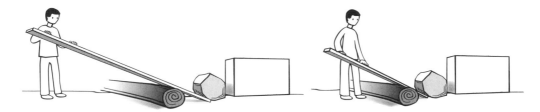

Alvin is trying to figure out how to lift a rock to place it on top of a block of cement. He has two strong, thick boards of different lengths. One board is twice as long as the second board. Alvin tries out both boards. First, he places the longer board over a log with the end of the board just under the rock as shown in the picture. He pushes down on the other end of the board. Then he does the same thing with the shorter board.

Part 1: Which sentence best describes the amount of *energy* it takes to lift the rock onto the top of the box? Circle your answer.

- **A** It is more for the long board.
- **B** It is more for the short board.
- **C** It is the same for both boards.

Part 2: Which sentence best describes the amount of *force* required to lift the rock onto the top of the box? Circle your answer.

- **A** It is more for the long board.
- **B** It is more for the short board.
- **C** It is the same for both boards.

Explain your thinking about how the lengths of the boards affect the amount of energy and force needed to lift the rock.

_____
_____
_____

Uncovering Student Ideas in Physical Science

Lifting a Rock

Teacher Notes

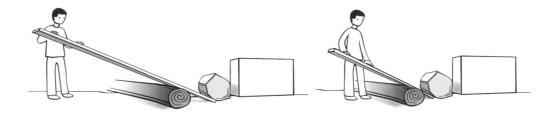

Purpose
The purpose of this assessment probe is to elicit students' ideas about levers and fulcrums. The probe is specifically designed to determine whether students can differentiate between the concept of force and the concept of energy in the context of levers and if they recognize how simple levers work.

Related Concepts
energy, fulcrum, lever, mechanical advantage, potential energy, simple machine, work

Explanation
The best answer to the question in Part 1 is C: It is the same for both boards. Because the object is moved by the same amount (from the floor to the top of the box), the total energy required in both cases is the same. The energy required to lift the rock is equal to the work done by the person pushing on the lever. This work is the force the person applies times the distance that he or she pushes on the lever. Although less force is required for the longer lever, the distance the force is applied is also longer. Another way to look at this is to compare the increase in gravitational potential energy of the rock. In both cases, the rock weighs the same and the distance it moves is the same. Therefore the change in gravitational potential energy is the same for both rocks.

The best answer to the question in Part 2 is B: It is more for the short board. The force needed for the longer lever arm is less than the force required to lift the rock with the shorter lever arm. It is the force times the distance from the fulcrum that is the important quantity for lifting. To reconcile the two answers (in Part 1 and Part 2), students will need to recognize that the energy required is the force times the distance that the force is acting. The longer arm requires less force, but the distance that the longer lever must move is also greater.

Administering the Probe
This probe is best used with middle school and high school students. Make sure that students can interpret the diagrams and that they explain their answers in terms of both energy and force.

Related Ideas in *National Science Education Standards* (NRC 1996) and *Benchmarks for Science Literacy* (AAAS 1993, 2009)
This probe and the other probes in this section (probes #39–#45) do not explicitly target key ideas in the national standards documents. Both the *National Science Education Standards* and *Benchmarks for Science Literacy* deliberately did not include simple machines as important ideas for science literacy. However, because most high school physics courses, many state and local standards, and National Science Foundation–funded reform curricula such as *Interactions in Physical Science; FOSS Simple Machines;* and STC's *Energy, Machines, and Motion* include levers as an extension of force, motion, and energy ideas, we decided it was important to include this "Other Topics" probe. In addition, students can use this probe to make a prediction before applying the technological design/engineering process described in the standards to solve design problems without necessarily knowing the specific content related to mechanical advantage, pulleys, and tension.

Related Research
- In spite of having a fairly good notion about the applications of a lever, many students have difficulty communicating these applications (Stepans 2008).
- The task of balancing is similar to the principles of levers. The related concept in physics is called torque, which is the force applied at a distance measured from the fulcrum. In Ortiz's research (Ortiz, Heron, and Shaffer 2005), she found that if students had not yet fully compiled their ideas about and procedures for using torque, they might see problems such as those in this probe as easy and obvious and thus might find it unnecessary to activate a physics principle. They would rely on a simple, intuitive explanation to make sense of the probe.

Suggestions for Instruction and Assessment
- Once students understand how forces and energy affect the motion of objects, provide them with experiences that enable them to apply this new information to thinking about how forces and energy are also involved in the way simple machines work. Make sure they first understand that a simple machine, such as a lever, is a simple device that affects the force required to perform a given task. This concept can later be connected to the concept of an input force (longer side of the lever) and an output force (shorter side of the lever).
- A common instructional strategy is to have students construct Rube Goldberg machines—a complex set of mechanical devices (using pulley and levers) built to complete a simple task (Jarrard 2008).
- In a Japanese Lesson Study, teachers designed a lesson that challenges students to lift heavy loads. The scenario is an earthquake and one of their friends becomes trapped beneath a heavy object. After planning what they might do, they are told by the teacher that they can use levers and pulleys to lift various loads. The title of the task is "Can you lift 100 kilograms?" See *www.lessonresearch.net/canyoulift1.html*.

References
American Association for the Advancement of Science (AAAS). 1993. *Benchmarks for science literacy.* New York: Oxford University Press.

American Association for the Advancement of Science (AAAS). 2009. Benchmarks for science literacy online. *www.project2061.org/publications/bsl/online*

Jarrard, A. 2008. The thinking machine: A physical science project. *Science Scope* (Nov.): 24–28.

National Research Council (NRC). 1996. *National science education standards.* Washington, DC: National Academies Press.

Ortiz, L., P. Heron, and P. Shaffer. 2005. Investigating student understanding of static equilibrium and accounting for balancing, *American Journal of Physics* 73 (6): 545–553.

Stepans, J. 2008. *Targeting students' physical science misconceptions using the conceptual change model.* Saint Cloud, MN: Saiwood Publications.

Mass, Weight, Gravity, and Other Topics

The Swinging Pendulum

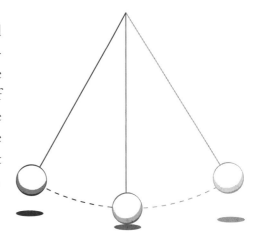

Gusti made a pendulum by tying a string to a small bob. He pulled the bob back and counted the number of swings the pendulum made in 30 seconds. He wondered what he could do to increase the number of swings made by the pendulum. If Gusti can change only one thing to make the pendulum swing more times in 30 seconds, what should he do? Circle what you think will make the pendulum swing more times.

A Lengthen the string.

B Shorten the string.

C Change to a heavier bob.

D Change to a lighter bob.

E Pull the bob back farther.

F Don't pull the bob back as far.

G None of the above. All pendulums swing the same number of times.

Explain your thinking. What rule or reasoning did you use to select your answer?

_____
_____
_____
_____
_____
_____

Uncovering Student Ideas in Physical Science

The Swinging Pendulum

Teacher Notes

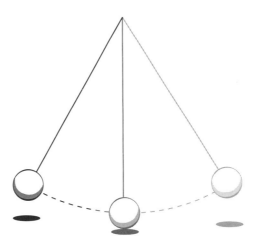

Purpose
The purpose of this assessment probe is to elicit students' ideas about pendulums. The probe is specifically designed to find out what variables students think affect the time it takes a pendulum to swing back and forth. In addition, if students have an opportunity to test the predictions they make for this probe, the probe can also be used to determine whether students recognize the need to control all but one variable.

Related Concepts
pendulum, periodic motion, variables

Explanation
The best answer is B: Shorten the string. Adding more weight (or mass) does not make a difference for the same reason that two objects that weigh differently will fall with the same acceleration. (See the Teacher Notes for probe #36, "Free-Falling Objects," on pp. 168–170). The initial height of the swing also does not make a difference for angles that are relatively small (less than about 40 degrees). This is because the higher the swing the more distance the bob has to travel, but the bob is also moving faster. If the bob is released from a small angle, then it will move slower, but does not have to travel as far. (*Note:* In the study of motion of pendulums, it is assumed, even with older students, that the angle of release is never above about 40 degrees. Larger-angle pendulums will behave differently because the period of motion is no longer a constant.)

Students who are learning about pendulums for the first time will discover through experiment that the period of a pendulum (the time it takes for a bob to swing back to its point of release) depends only on the length—that is, the distance from the point of support to the center of mass (also called center of gravity) of the bob. If students think about air resistance acting on the bob, then they may also select C (change to a heavier bob). However, in most cases, students who select C are not thinking about air resistance when they choose this option (you will need to check their reasoning). (*Note:* The dependence of the period on the length of the string (or wire) is true only (a) for small angles, i.e., less than about 40 degrees, and (b) if the mass of the bob is much greater than the mass of the wire.)

Administering the Probe

This probe can be used with upper elementary and middle school students. Show students a pendulum, point out what the bob is, and demonstrate its swinging motion. With older students, consider referring to the mass of the pendulum bob instead of using the words *heavier* and *lighter* that now appear in the probe's distracters.

Related Ideas in *National Science Education Standards* (NRC 1996) and *Benchmarks for Science Literacy* (AAAS 1993, 2009)

This probe and the other probes in this section (probes #39–#45) do not explicitly target key ideas in the national standards documents. Both the *National Science Education Standards* and *Benchmarks for Science Literacy* deliberately did not include periodic motion as an important idea for science literacy. However, because simple pendulums are frequently used with elementary and middle school students to build on the notion of a "fair test" by introducing and practicing the skill of identifying and controlling variables, we decided it was important to include this "Other Topics" probe.

Related Research

- Students often think mass or weight is the primary factor affecting the period of a pendulum. Some think a pendulum with a lighter bob moves faster while others think that a pendulum with a heavier bob moves faster (Stepans 2008).

- Some students cannot distinguish the effects of gravity, air resistance, and friction from factors that affect the period of a pendulum (Stepans 2008).

- In a study by Carey et al. (1989) upper elementary and middle school students had difficulty understanding experimentation as a method of testing ideas. They tended to view experimentation as a method of trying things out or producing a desired outcome (AAAS 1993).

- Students of all ages may overlook the need to hold all but one variable constant (AAAS 1993).

- Although young children have a sense of what it means to run a fair test, they frequently cannot identify all of the important variables, and they are more likely to control those variables that they believe will affect the result. The more familiar students are with the topic of a given experiment the more likely they are to identify and control variables (AAAS 1993).

Suggestions for Instruction and Assessment

- This probe can be used to launch into an experiment where students need to identify and control different variables in order to determine which factor affects the time it takes for a pendulum to swing.

- With younger children, stress the need to conduct a "fair test." Ask students what needs to be kept the same in order to make the testing of their different ideas "fair." The notion of a fair test with younger children is a precursor to developing an understanding of variables and controls in later, more sophisticated experiments.

- This probe can be combined with "Grandfather's Clock," an everyday science mystery story (Konicek-Moran 2008) that helps students discover how lengthening or shortening a pendulum helps a clock keep time.

- It is not just the length of string that always determines the period (the time it takes to complete one back and forth swing). Help students be aware that it is possible to have the same string length and the same bob or different bobs with the same mass, yet have different periods. The period depends on the shape and orientation of the bob attached to the string. For example, a block shaped like  could have a string tied around its center (hung horizontally) or suspended from its top (hung vertically). When hung vertically, its center of mass (or center of gravity) is lower and the string length would have to be adjusted to account for that. If the string lengths were kept the same, the bob in the horizontal orientation would swing faster than the bob in the vertical orientation because their centers of mass are different. The length of a pendulum is always measured from the end of the string to the center of mass of the bob.
- With young children, avoid the use of terms like *period* and *frequency*. Instead ask them which pendulum swings more times in a given time interval (e.g., 30 seconds or one minute).

References

American Association for the Advancement of Science (AAAS). 1993. *Benchmarks for science literacy.* New York: Oxford University Press.

American Association for the Advancement of Science (AAAS). 2009. Benchmarks for science literacy online. *www.project2061.org/publications/bsl/online*

Carey, S., R. Evans, M. Honda, E. Jay, and C. Unger. 1989. An experiment is when you try it and see if it works: A study of grade 7 students' understanding of the construction of scientific knowledge. *International Journal of Science Education* 11: 514–529.

Konicek-Moran, R. 2008. *Everyday science mysteries: Stories for inquiry-based teaching.* Arlington, VA: NSTA Press.

National Research Council (NRC). 1996. *National science education standards.* Washington, DC: National Academies Press.

Stepans, J. 2008. *Targeting students' physical science misconceptions using the conceptual change model.* Saint Cloud, MN: Saiwood Publications.

Mass, Weight, Gravity, and Other Topics

Bicycle Gears

Maureen is building a bicycle. She has four gears: two large and two small, as shown in pictures A, B, C, and D. She wants to be able to ride the bicycle up steep hills without having to pedal too hard. She needs help figuring out what the best gear size is for the front and back wheels and where to attach the pedals. Which sentence best describes how she should build her bicycle?

A Put the pedals on the big gear and the small gear on the back wheel.

C Put the big gears on both the front and the back wheels.

B Put the small gear on the pedals and the large gear on the back wheel.

D Put small gears on both the front and back wheels.

Explain your thinking. Describe how you decided which gears to use and where to place them.

_____
_____
_____
_____
_____

Uncovering Student Ideas in Physical Science   205

Bicycle Gears

Teacher Notes

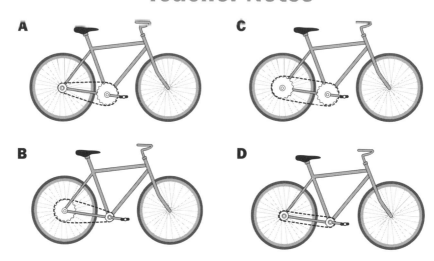

Purpose
The purpose of this assessment probe is to elicit students' ideas about gears. The probe is specifically designed to determine whether students use the concept of gear ratios.

Related Concepts
energy, gears, gear ratio, work

Explanation
The best answer is B: Put the small gear on the pedals and the large gear on the back wheel. The turning of one gear transfers energy to the other gear. Gears behave like levers—the larger the gear that is turned, the easier it is to turn the smaller gear.

The pedals should be on the small gear and the wheel should be on the large gear in order to ride the bicycle uphill. This is because you want one complete turn of the pedals to move the bicycle a short distance. The shorter the distance the bicycle moves with each turn of the pedal, the less force it requires from the rider (for each complete rotation of the pedals). To go fast, you would want to switch the gears and pedal with the large gear and mount the wheel on the small gear.

Administering the Probe
This probe is best used with high school students. Students may need to have some familiarity with riding bicycles and changing gears before the probe is administered. Make sure students understand that the chain is very light and therefore the mass is not a factor in this problem.

Related Ideas in *National Science Education Standards* (NRC 1996) and *Benchmarks for Science Literacy* (AAAS 1993, 2009)
This probe and the other probes in this section (probes #39–#45) do not explicitly target key ideas in the national standards documents. Both the *National Science Education Standards*

Mass, Weight, Gravity, and Other Topics

and *Benchmarks for Science Literacy* deliberately did not include simple machines or the elements that make them up as important ideas for science literacy. Because many high school physics curricula include elements of simple machines—such as gears and ball bearings—as an extension of force, motion, energy, and work ideas, we decided it was important to include this "Other Topics" probe. In addition, students can use this probe to make a prediction before applying the technological design/engineering process described in the standards to solve design problems without necessarily knowing the specific content related to gears, gear ratios, and work. Both *Benchmarks for Science Literacy* and *Principles and Standards for School Mathematics* (NCTM 2000) target ratios as important learning goals. This probe is an example of using ratio in the context of a technological design problem.

Related Research

- The standards of the National Council of Teachers of Mathematics call for teaching ratios and ratio reasoning in the middle school years. However, research indicates that students need to be exposed to these ideas at a younger age if they are ultimately to be successful with ratios and ratio reasoning (NCTM 2000).
- Researchers found that there was little correlation between a child's computational ability and his or her ability to reason with ratios or proportions. The researchers suggest that young students be given multiple opportunities to use ratios in a wide variety of contexts (Misailidou and Williams 2003).

Suggestions for Instruction and Assessment

- An example of an electronic resource for the mathematics standard on ratios (related to this probe) is the computer simulation on the NCTM website on bicycle and gear ratios. This collection of resources is from NCTM's work called "Illuminations." See *http://illuminations.nctm.org/ActivityDetail.aspx?ID=178*.
- With older students, bicycle wheels and gearing ratios can be used to teach about exponential functions by analyzing a multigear bicycle (such as a 10-speed bike). For a description of this activity see Greenslade (1979).

References

American Association for the Advancement of Science (AAAS). 1993. *Benchmarks for science literacy.* New York: Oxford University Press.

American Association for the Advancement of Science (AAAS). 2009. Benchmarks for science literacy online. *www.project2061.org/publications/bsl/online*

Greenslade, T. 1979. Exponential bicycle gearing. *The Physics Teacher* 17: 455.

Misailidou, C., and J. Williams. 2003. Diagnostic assessment of children's proportional reasoning. *Journal of Mathematical Behavior* 22: 335–368.

National Council of Teachers of Mathematics (NCTM). 2000. *Principles and standards for school mathematics.* Reston, VA: NCTM.

National Research Council (NRC). 1996. *National science education standards.* Washington, DC: National Academies Press.

Index

Note: Page numbers in *italics* refer to charts.

A

Acceleration, 2, 20–21, 48, 51, 52–53, 56, 116, 168
   decelerating as slowing down, 57
   of free-falling objects, 168
   going faster equated with, 57
   misconceptions about, 57, 105, 117–118, 121, 169–170
   national standards and, 53
   negative, *slowing down* and, 57–58
   velocity and, 58
   See also Motion; Speed
Action-reaction pairs, 129
Active action, 80, 108, 109
Activity Before Concept (ABC) approach, 53
Air, fluid friction, 85
Airplanes, forces on apples in, 107–108, 110
Animals, height data, 65
Apples
   in airplanes, forces on, 107–108, 110
   falling from trees, gravity and, 163–164
Aristotelian thinking, 3, 80
Arons, Arnold, 2
Ausubel, David, 8
Average speed, 40, 41, 48

B

Balancing, 189–191, 194
   balance beams, 193–194
   levers and, 199
   making small balance beams, 195–196
   symmetrical and nonsymmetrical objects, 195
   torque and, 199
   unifix cubes, 195
Balls
   dropping from a tower, 177–178
   pushing around a corner, 137
   released from a merry-go-round, 114
   rolling speeds of, 43–44, 48–49
   throwing while riding a skateboard, 101
   twirling on a string, 111–112
Bernoulli's law, 155
Bicycle gears, 205–206, 207
Birds, gravity and, 160
Brachistochrones, 62
Bridging analogies, 73–74, 82, 85, 109–110, 180
Buoyant force, 150

C

Card sorting
   as FACT (formative assessment classroom technique), 36–37
   for gravitational force lesson, 159
Cars
   forces when riding in, 135–136
   toy car travel measurement, 15–16
Center of mass, 190, 191
Change
   computing rates of, 49
   in motion, calculating magnitude of, 117
Changing direction, 104
   velocity and, 56
Circular motion, 111–112, 113, 136
   misconceptions about, 113
   riding in a car, 135–136, 137
Circular objects
   circumference, 64
   string around the Earth exercise, 63–64
Circumference, 64
   pi and, 64
Clock readings, 32, 36
   ambiguity with time, 38
   pendulums' effect on, 203
Computational ability
   proportional reasoning and, 37, 65, 207
   using ratios and, 207
Concept matrices, xii
   forces and Newton's laws, *68*
   mass, weight, and gravity, *140*
   motion and position, *12*
Conceptual learning, ix
Conservation of energy, 93
Conservation of mass, 144, 146
Constant motion, 2
Constant speed, 20, 21, 24, 28, 32, 80, 89
   moving on a parade float and, 99–100
   pulling an object upward and, 124
   of rolling ball, 43–44
   unbalanced force and, 45, 81, 97
   See also Speed
Contact force, 84

D

Delta symbol, use in textbooks, 38

Index

Diagnoser project, 2, 49
Displacement, 36, 44
   speed vs., 38
Distance, 24, 28
   as distance traveled, 34
   travel measurement, 15–16

E
Earth
   as a sphere, 179, 180
   gravitational field, 174
Electricity, charged objects, 73,77
Energy, 206
   conservation of, 93
   potential, 174
   student use of the word, 45–46
   studying before learning about motion, 82
   work, 186, 206
Energy transfer, 92, 93
Energy transformations, 93
Experimentation, misconceptions about, 203

F
FACT (Formative Assessment Classroom Technique)
   card sorting for, 36–37
   Misrepresentation Analysis, 105
   See also Formative assessment
Falling objects, 115–116
   misconceptions about, 179
   teaching difficulties, 4–5
Finger strength, 127–128
Footracing, speed and, 47–48
Force
   acting along a straight line, 93
   action-at-a-distance force, 76, 78
   on apples in airplanes, 107–108, 110
   centrifugal, 137
   centripetal, 112, 113
   concept matrix, *68*
   contact force, 76, 108
   equal forces, 128, 129, 131–133
   falling objects and, 169
   forces of nature, 160, 165
   friction as, 85
   "holding force," 72
   magnetic, 84
   misconceptions about, 45–46, 77, 80, 81, 92, 93, 97, 105, 109, 121, 129, 137
   motion and, 33, 45, 46, 79–80, 93, 117, 121, 183
   net force, 81, 85, 93, 113, 119–120
   normal, 108
   between objects, 128
   passive support and, 109
   preconceptions about, 2
   as property of an object, 81
   resting state and, 25, 73, 93, 109
   teaching difficulties, 3–4
   touch as necessary for, 75–76
   transfer between objects, 81
   types of, 76
   using spring scales to indicate, 125
   weight and, 186
   when riding in cars, 135–136
   whirling objects on a string, 111–112, 113
   See also Gravitational force; Pushes and pulls; Unbalanced force
Force and Motion Conceptual Evaluation, 2
Force Concept Inventory, 2, 113
Forces of nature, 160, 165, 173, 179
Formative assessment, ix–x
   See also FACT
Free fall, 167–168, 168
   acceleration and, 168
   duration of fall, 168
   force and, 169
   misconceptions about, 169
   weight and, 169
   See also Gravitational force
Frequency, 204
Friction, 83–84, 88, 92
   effect on magnets, 84
   electric forces and, 85
   fluid, 85
   force and, 85
   interaction and, 85
   kinetic friction, 84
   low-friction devices, 97
   misconceptions about, 85, 97
   in moving vs. stationary objects, 85
   on playground slide, 83–84
   rolling friction, 84, 93, 120
   as rubbing, 85, 89
   sliding friction, 84, 93
   static friction, 84
   torque and, 120
   world without friction exercise, 87–88
Fulcrums, 195

G
Galileo, 80
Gases, friction and, 85
Gear ratio, 206
Gears, 206
Going faster ambiguity, 21, 25, 57, 61
Graphs, 25, 27, 28, 32
   ambiguities in textbooks, 34
   changes in circumference and diameter, 66
   distance/time, 27, 29
   interpretation, pictorial vs. mathematical, 32
   interpretation difficulties, 29, 33

Index

as literal pictures, 29, 33
of miles per hour, 37
motion diagrams as, 27–28, 32
position vs. time, 27, 49
as symbolic relationships, 29
Gravitational force, 73, 77, 108, 109, 116, 150, 158, 159, 164–167, 172, 178, 179
    apples falling from trees and, 163–164
    applying to real-life concepts, 161
    astronauts' Moon boots and, 158
    on buried objects, 174
    decrease with distance from earth's surface, 173
    defined, 165
    direction of fall on Earth's surface, 179
    distance from Earth's surface and, 172
    experiencing, 157–158
    heavy objects and, 117, 118, 169
    instructional difficulties, 3
    mass and, 158–159
    misconceptions about, 109, 117–118, 151, 160, 161, 165, 173–174, 179
    modeling with rubber bands, 161, 166
    on the Moon, 144–145, 158, 160
    moving objects and, 159
    pressure and, 155
    as pulling force, 72
    Shell Theorem, 174
    sizes of objects and, 158
    slowed motion and, 45, 46
    speed and, 159
    strength of, 173
    students' prior learning about, importance of, 174
    weight and, 144–145, 155, 158
    See also Force; Free fall
Gravitational potential energy, 172, 174
Gravity, 157–159, 164, 168, 172, 178
    concept matrix, *140*
    Curriculum Topic Study (CTS) chart for probe development, *xxi*
    instructional difficulties, 3
    nature of, 3
    planetary orbits and, 77, 81, 159, 165, 173
    rock location and, 171–172
Gunstone, Richard, 1
Gut dynamics, 101

H
Harrington, Rand, xxvi

I
Ice cream drip exercises, 23, 27
Imagination, learning scientific principles and, 89
Inertia, 97
Instantaneous speed, 48
Instructional difficulties, 8

difference between weight and mass, 3–4
emerging concepts and language, 5–6
force and motion probes, 7–8
teaching about gravity, 3
understanding motion of falling objects, 4–5
Interaction, 80, 84, 92, 108, 128
    equal and opposite reaction, 128, 129, 131–133
    between forces, identifying, 133
    friction and, 85
    between inanimate objects, 107–108, 110
    stretching a rubber band, 128
Internet sources. *See* Websites

J
Jogger and Sprinter Elicitation (Internet feature), 49

K
K-W-L (Know-Want to know-Learned) strategy, 196
Karplus, Robert, 2
Keeley, Page, xxv
Kinematics, 45, 82

L
Lay dynamics, 101
Learning
    imagination role in, 89
    studying energy before motion, 82
Levers
    balancing and, 199
    length of, 197, 198
    lifting heavy loads with, 183, 187, 197–198
    using Rube Goldberg machines to demonstrate, 183, 187
Lifting heavy loads, 185–186, 187
Lines, straight confused with horizontal, 112
Logs, balance points for cutting, 189–191

M
Magnetism, students' concepts of, 77
Magnets, 73, 77
    friction effect on, 84
Mass, 144, 194
    bulk appearance and, 146
    center of, 190, 191
    concept matrix, *140*
    defined, 145
    massive confused with, 145–146, 191
    misconceptions about, 146–147, 191
    weight vs., 146
Measurement
    of object's position over time, 25
    student ethnicity and, 17
    See also Units of measurement
Measurement scales, starting points, 16, 17
Mechanical advantage, 182, 186

Index

Mechanics Baseline Test, 2
Misconceptions
   acceleration, 57, 105, 117–118, 121, 169–170
   changing incorrect beliefs, 121
   circular motion, 113
   experimentation, 203
   falling objects, 179
   force, 45–46, 77, 80, 81, 92, 93, 97, 105, 109, 121, 129, 137
   friction, 85, 97
   gravitational force, 109, 117–118, 151, 160, 161, 165, 173–174, 179
   gravity in water, 151
   language and, 5–6
   mass, 146–147, 191
   motion, 97, 105, 113, 121, 125, 129, 133
   outer space, 118, 160
   resting objects, 73, 85, 109, 166
   unbalanced force, 125
   weight, 169
   See also Preconceptions
Mobiles, to demonstrate balance, 195
Moments, 195
Momentum, student use of the word, 45–46
Moon, gravity on the, 144–145, 158, 160
More A therefore More B concept, 41, 49, 57, 183
Motion, 49
   on a parade float, 99–100
   calculating change in, 117, 169
   change in, 2
   circular, 111–112, 113
   comparisons of, 48
   constant, 89
   describing, language tools for, 53, 57
   direction of, 21, 45, 55–56, 80, 89, 113
   force and, 33, 45, 46, 79–80, 93, 117, 183
   learning about energy first, 82
   misconceptions about, 97, 105, 113, 121, 125, 129, 133
   in outer space, 95–96
   periodic, 202
   position and, 21
   concept matrix, *12*
   position vs. time graph, 49
   rates of change in, 49
   relative, 49, 99–100
   roller coasters, 55–56
   slowing of, gravity and, 46
   speed and, 25, 113
   stopping when force runs out, 73, 81, 89, 97
   teaching difficulties, 3–4
   unbalanced force and, 45, 81, 113
   uniform, 20, 24, 28
   See also Acceleration; Speed
Motion detectors, 21, 33
Motion diagrams, 24–25
   creating, 25
   graphs as, 27–28
   ice cream drip exercise, 23
   interpreting intervals, 24
   of rolling balls' speeds, 48–49
Movement direction
   changes in, 56, 104
   speed and, 21
   types of, 25

N

National Science Teachers Association (NSTA)
   Learning Center resources, xv, 13, 69–70, 141–142, 207
   publications, xv, 13, 69, 141
   website, xv
National Standards, xiii–xiv
   CTS guides and, xix
Nature. *See* Forces of nature
Net force, 81, 85, 93, 113, 119–120
Newton, Isaac, 80
Newtonian thinking, 3
Newton's first law of motion, 88, 89, 96, 100, 104, 105, 112, 136
Newton's laws of motion, xviii, 155
   concept matrix, *68*
Newton's second law of motion, 58, 80, 105, 119–120, 124, 168
Newton's third law of motion, 105, 128, 129–130, 131–132, 133
   misconceptions about, 129, 133
   phrasing of, 130, 133–134
   springs as an example of, 133
Newton's universal law of gravity, 158

O

Outer space
   misconceptions about, 118, 160
   motion in, 95–96

P

P-E-O-E (Predict-Explain-Observe-Explain) strategy, 44–45, 61, 114, 121, 151, 192
P-E-O (Predict-Explain-Observe) strategy, 44–45, 61, 195
P-prims, 183
Parades, riding on a float in, 99–100
Passive action, 80, 108
Passive support, force and, 109
Pendulums, 201–203
   bob weight, 203
   clock time and, 203
Periodic motion, 202
   pendulum length and, 203, 204

Index

Physics Cinema Classics (film), 101
Pi, 64
Piaget, Jean, 1, 155
Pizza dough, weight of, 143–144
Position, 24, 28, 32, 36, 48
   measurement points and, 16, 17
   motion and
      concept matrix, *12*
      push or pull strength effect, 72
   as an object's location, 34
   time vs., 27–29
Potential energy, 174
Preconceptions, xi
   about force, 2
   as springboards for learning, xi
   See also Misconceptions
Pressure, 154–155
   Bernoulli's law, 155
   walking on eggs, 156
   walking on snow, 156
   weather and, 155
   weight and, 154
Probes, ix–x
   considerations in using, xv
   force and motion, 6–7
   in grades 3–5, xv
   in grades K-2, xv–xvi
   in high school, xvii–xviii
   in middle school, xvi–xvii
   ways to administer, xiii
Proportion, 64
Proportional reasoning, 29, 37, 41
   algorithms and, 65
   computational ability and, 37, 65, 207
   influences on difficulty with, 65
Pulleys, 182, 186
   size factors, 181–182
   using Rube Goldberg machines to demonstrate, 183, 187, 199
Pushes and pulls, 71–72, 81, 89, 136, 164, 165
   action at a distance and, 78
   bridging analogies, 73–74
   constant speed and, 124
   electrically charged objects and, 73, 77
   kicking as, 73
   lifting a bucket, 123–124
   by magnets, 73
   pulling a person up a cliff, 185–186
   pulling a spool of string, 119–120
   throwing as, 73
   on toy traveling in a straight line, 114
   See also Force

Q
Quantity vs. change in quantity, difficulty in differentiating, 2

R
Racing
   on foot, 47–48
   Jogger and Sprinter Elicitation, 49
   NASCAR, 51–52
Ramps
   bank curvature and, 62
   roller coasters, 55–56
   skateboarding on, 19–20
   speed on, 59–60
Ratios, 36, 40, 64
   circumference to diameter, 64
   computational ability and, 207
   differing difficulties of, 37, 41
   gear ratios, 206
   miles per hour (mph) as, 37, 40
   speed vs. exchange, 37, 65
   teaching about, 207
Rest, misconceptions about, 73, 85, 109, 166
Roller coasters, 55–56
Roller skates, 91–92
Rolling friction, 84, 93, 120
Rolling objects, 43–44, 48–49, 59–60, 114, 137
Round objects. *See* Circular objects
Rubber bands
   for interaction demonstration, 128
   for studying gravitational force, 161, 166
Rube Goldberg machines, to demonstrate pulleys and levers, 183, 187, 199

S
Scales
   to indicate applied force, 125
   to study equal forces, 129
Science Curriculum Topic Study (Keeley), xix–xx
Science Formative Assessment (Keeley), xviii–xix
Scientific Terminology Inventory Probe (STIP), 53, 56
Simple machines, 182, 183, 186, 199
Skateboards
   motion and speed of, 19–20
   throwing a ball while riding, 101
Snow, walking on, 156
Space. *See* Outer space
Spaceships, design of, 103–104, 105
Spatial reasoning, 33
Spatial reasoning ability, graph interpretation and, 33
Speed, 24, 28, 32, 36, 44, 48, 52, 56
   average, 40, 41, 48
      calculation of, 52
   comparing racers' speeds, 47–49
   defined, 64

Index

delta symbol for representing, 38
direction of object's movement and, 21
displacement vs., 38
going faster ambiguity, 21, 25, 57, 61
gravitational force and, 159
instantaneous, 48
larger measurement units for greater speed, 41
measurement units for, 39–40, 41
miles per hour, 37, 40
as ratios, 38, 39–40
snapshot descriptions, 21, 61
velocity vs., 51, 52
See also Acceleration; Constant speed; Motion
Speedometers, determining accuracy of, 35–36
Spherical Earth, 178
Sports, lay dynamics, 101
Standards. *See* National Standards
STIP. *See* Scientific Terminology Inventory Probe
Students, conceptual learning and, ix
Students' misconceptions. *See* Misconceptions
Surface deformation, 109–110
Symbolic relationships, graphs as, 29

T

Teaching. *See* Instructional difficulties
Tension, 120, 182, 186
Textbook ambiguities
 in distance/time graphs, 33–34
 distance vs. displacement, 38
 distance vs. position, 38
 quantity vs. change in quantity, 37–38
 speed vs. displacement, 38
 time vs. clock reading, 38
Ticker-Tape Puzzle (reasoning task), 25
Time, 24, 28, 32
 clock reading ambiguity with, 38
Time intervals, 24, 28, 32, 36, 44, 48
 short and *long* as, 61
Torque, 189–191, 199
 friction and, 120
Touch, defined, 76
Turning effect, 189

U

Unbalanced force, 45, 81, 85, 89
 constant speed and, 45, 105, 124, 125
 misconceptions about, 125
 motion and, 45, 81, 113, 124
 See also Force
Uncovering Student Ideas in Science series (Keeley and Harrington), ix–x, xi–xii
Uniform motion, 20, 24, 28, 32, 44
Units of measurement, 40, 41
 conversion tables, 41
 determining appropriate units, 41
 larger units for greater speed, 41

miles per hour, 37, 40
for speed, 39–40, 41
See also Measurement

V

Variables, 202
 constant, 203
 identifying which are important, 203
Vector, concept of, 53, 58
Velocity, 52, 53, 56
 acceleration and, 58
 change of direction and, 56
 constant velocity, 52
 speed vs., 51

W

Water
 direction from garden hoses, 114
 fluid friction, 85
 ice cubes in, 151
 misconceptions about gravity in, 151
 volume vs. weight, 155–156
 weight of, 153–154
 weight of objects in, 149–150
Websites
 on bicycle and gear ratios, 207
 Blick on Flicks (film reviews for scientific accuracy), 105
 "Can you lift 100 kilograms" exercise, 199
 CD hovercraft (low-friction device), 97
 on curved banks and rolling objects, 62
 diagnoser.com, 49
 on force and motion, 69–70
 NSTA, xv
 NSTA Learning Center Resources, xv, 13, 69–70, 141–142, 207
 Physics Cinema Classics (film), 101
 "string around the Earth problem" search, 65
Weight, 73, 77, 108, 109, 116, 144, 150, 154, 189, 194
 in and out of water compared, 149–150
 concept matrix, *140*
 English vs. metric measurement, 146
 "felt" weight, 145, 146, 155
 force and, 186
 free fall and, 169
 gravitational force and, 144–145, 155
 mass vs., 146
 misconceptions about, 169
 object placement and, 155
 of pendulum bobs, 203
 of pizza dough, 143–144
 pressure and, 154
 as pulling down force, 146
 of water, 153–154, 155
White, Richard, 1
Work (energy), 186, 206